OPTIMAL ESTIMATION OF PARAMETERS

This book presents a comprehensive and consistent theory of estimation. The framework described leads naturally to the maximum capacity estimator as a generalization of the maximum likelihood estimator. This approach allows the optimal estimation of real-valued parameters, their number and intervals, as well as providing common ground for explaining the power of these estimators.

Beginning with a review of coding and the key properties of information, the author goes on to discuss the techniques of estimation, and develops the generalized maximum capacity estimator, based on a new form of Shannon's mutual information and channel capacity. Applications of this powerful technique in hypothesis testing and denoising are described in detail.

Offering an original and thought-provoking perspective on estimation theory, Jorma Rissanen's book is of interest to graduate students and researchers in the fields of information theory, probability and statistics, econometrics, and finance.

JORMA RISSANEN was a member of research staff in the IBM Almaden Research Center from 1965 to 2001, and is currently Professor Emeritus at Tampere University of Technology, Finland. Among his main results are the introduction of the MDL principle for statistics, the invention of arithmetic coding, and the introduction of variable-length Markov chains with the associated Algorithm Context. He has received many awards, including the 2007 Kolmogorov Medal from the University of London's CLRC, and the 2009 Shannon Award from the Information Theory Society. He received two Outstanding Innovation Awards from IBM, one in 1980 and the other in 1988, and a Corporate Award in 1991.

The minimum description length (MDL) principle is a very universal principle of statistical modeling in estimation, prediction, testing, and coding. Jorma Rissanen, the pioneer of the MDL principle, evolves a new theory to reach the most general and complete notion, which he calls the complete MDL principle. In this book the author derives it by introducing the key notion of maximum capacity. The most fundamental methods of estimation such as maximum likelihood estimation and the MDL estimation are naturally derived as the maximum capacity estimators, and their optimality is justified within a unifying theoretical framework. Through the book, readers can revisit the meaning of estimation from the author's very original viewpoint, and will enjoy the most advanced version of the MDL principle.

Kenji Yamanishi, *The University of Tokyo*

In this splendid new book, Jorma Rissanen, the originator of the minimum description length (MDL) Principle, puts forward a comprehensive theory of estimation which differs in several ways from the standard Bayesian and frequentist approaches. During the development of MDL over the last 30 years, it gradually emerged that MDL could be viewed, informally, as a maximum *probability* principle that directly extends Fisher's classical maximum *likelihood* method to allow for estimation of a model's structural properties. Yet providing a formal link between MDL and maximum probability remained elusive until the arrival of this book. By making the connection mathematically precise, Rissanen now ties up the loose ends of MDL theory and at the same time develops a beautiful, unified, entirely original and fully coherent theory of estimation, which includes hypothesis testing as a special case.

Peter Grünwald, *Centrum voor Wiskunde en Informatica, The Netherlands*

OPTIMAL ESTIMATION OF PARAMETERS

JORMA RISSANEN
Tampere University of Technology
Helsinki Institute for Information Technology

CAMBRIDGE UNIVERSITY PRESS
Cambridge, New York, Melbourne, Madrid, Cape Town,
Singapore, São Paulo, Delhi, Mexico City

Cambridge University Press
The Edinburgh Building, Cambridge CB2 8RU, UK

Published in the United States of America by Cambridge University Press, New York

www.cambridge.org
Information on this title: www.cambridge.org/9781107004740

© Cambridge University Press 2012

This publication is in copyright. Subject to statutory exception
and to the provisions of relevant collective licensing agreements,
no reproduction of any part may take place without the written
permission of Cambridge University Press.

First published 2012

Printed in the United Kingdom at the University Press, Cambridge

A catalog record for this publication is available from the British Library

Library of Congress Cataloging in Publication data
Rissanen, Jorma.
Optimal estimation of parameters / Jorma Rissanen, Tampere University of Technology,
Helsinki Institute for Information Technology.
    pages cm
ISBN 978-1-107-00474-0 (hardback)
1. Estimation theory.   I. Title.
QA276.8.R57 2012
519.5′44 – dc23   2012007503

ISBN 978-1-107-00474-0 Hardback

Cambridge University Press has no responsibility for the persistence or
accuracy of URLs for external or third-party internet websites referred to in
this publication, and does not guarantee that any content on such websites is,
or will remain, accurate or appropriate.

Contents

Preface		*page* vii
1	**Introduction**	1
Part I	**Coding and information**	9
2	**Basics of coding**	11
	2.1 Coding and probability	12
	2.2 Coding and entropy	15
	2.3 Huffman's code	16
	2.4 Codes as number representations	18
	2.4.1 Arithmetic codes	21
3	**Basics of information**	24
	3.1 Main definitions	24
	3.2 Mutual information	25
	3.2.1 Expected mutual information	25
	3.2.2 Maximum estimation information	27
Part II	**Estimation**	33
4	**Modeling problems**	35
	4.1 Models	36
	4.2 General comments on estimation	39
	4.3 Maximum capacity	41
	4.4 Necessary conditions for optimality	47
	4.5 General and complete MDL principles	51
5	**Other optimality properties**	57
	5.1 Minmax problems	57
	5.2 Consistency	63
6	**Interval estimation**	70
	6.1 Optimum intervals	71

		6.2	Maximum capacity partition	75
		6.3	Error probability	77
			6.3.1 Asymptotic distinguishability	78
			6.3.2 Partition algorithm for $k = 2$	81

7 Hypothesis testing — 83

- 7.1 General plan — 84
- 7.2 Test statistics — 87
- 7.3 Characteristic histograms — 88
- 7.4 Main tests — 92
  - 7.4.1 Simple hypothesis — 96
  - 7.4.2 Composite hypothesis — 98
  - 7.4.3 Likelihood ratio — 102

8 Denoising — 104

- 8.1 Hard thresholding — 105
- 8.2 Soft thresholding — 109

9 Sequential models — 112

- 9.1 Bernoulli class — 114
- 9.2 Variable order Markov chains — 116
  - 9.2.1 Algorithm Context — 119
  - 9.2.2 Tree pruning — 120
  - 9.2.3 Universal code — 123
  - 9.2.4 Extension to time series — 124
- 9.3 Linear quadratic regression models — 127
  - 9.3.1 Fixed variance — 129
  - 9.3.2 Free variance — 129
- 9.4 Autoregressive moving average (ARMA) models — 133
  - 9.4.1 Prediction — 134
  - 9.4.2 Prediction bound with estimated parameters — 141

Appendices

A Elements of algorithmic information — 144

- A.1 Recursive and partial recursive functions — 145
  - A.1.1 Universal computers — 147
  - A.1.2 Relative randomness and typicality — 149
- A.2 Kolmogorov structure function — 150

B Universal prior for integers — 153

References — 156
Index — 161

Preface

I have a long lasting interest in estimation, which started with attempts to control industrial processes. It did not take long to realize that the control part is easy if you knew the behavior of the process you want to control, which meant that the real problem is estimation. When I was asked by the Information Theory Society to give the 2009 Shannon Lecture I thought of giving a coherent survey of estimation theory. However, during the year given to prepare the talk I found that it was not possible, because there was no coherent theory of estimation. There was a collection of facts and results but they were isolated with little to connect them. To my surprise this applied even to the works of some of the greatest names in statistics, such as Fisher, Cramér, and Rao, which I had been familiar with for decades, but which I had never questioned until now that I was more or less forced to do so. As an example, the famous maximum likelihood estimator due to Fisher [12] had virtually no formal justification. The celebrated Cramér–Rao inequality gives it a non-asymptotic justification only for special models and for more general parametric models only an asymptotic justification. Clearly, no workable theory should be founded on asymptotic behavior. About the value of asymptotics, we quote Keynes' famous quip that "asymptotically we all shall be dead."

In trying to prepare the talk I was faced with a large jigsaw puzzle in which the pieces did not quite fit, and some crucial pieces were completely missing. After a considerable struggle I was able to make the pieces fit but to do so I had to alter virtually all of them and ignore the means and concepts introduced by the masters, who of course did not have access to information and coding theory. Their followers, however, did, but since this theory did not immediately solve their problems it was regarded as irrelevant and ignored. About the only concept that survived was the maximum likelihood estimator, and even that was as a special case of a more powerful *maximum capacity* estimator, which permits estimation of the number of parameters as well as intervals. The result is a theory of estimation which covers all the aspects of estimation that I can think of including, even hypothesis testing, which in this treatment is founded on estimation, and is quite different from the usual.

1 Introduction

This book has two parts: the first summarizes the facts of coding and information theory which are needed to understand the essence of estimation and statistics, and the second describes a new theory of estimation, which also covers a good part of statistics as well. After all, both estimation and statistics are about extracting information from the often chaotic looking data in order to learn what it is that makes the data behave the way they do. The first part together with an outline of the algorithmic information in Appendix A is meant for the statistician who wants to understand his or her discipline rather than just learn a bag of tricks with programs to apply them to various data, tricks that are not based on any theory and do not stand a critical examination although some of them can be quite useful, providing solutions for important statistical problems.

The word *information* has many meanings, two of which have been formalized by Shannon. The first is fundamental in communication as just the number of messages, strings of symbols, either to be stored or to be sent over some communication channel, the practical question being the size of the storage device needed or the time it takes to send them. The second meaning is the measure of the strength of the statistical property a string has, which is fundamental in statistics, and very different from that in communication. Shannon formalized the measure of information of both kinds in terms of the famous *entropy* of a probability distribution $P(x)$ on the strings:

$$H(P) = \sum_x P(x) \log 1/P(x). \tag{1.1}$$

This is an abstract and intricate measure, for it does not require interpretation of the nature of the strings and symbols x, as done in the thermodynamic entropy in physics. It plays a central role in coding, for, as we shall discuss in Section 2.2, it amounts to the greatest lower bound of the mean code length of strings emitted by the *source P*, as the jargon goes. It generalizes Hartley's measure as the logarithm of the number of elements in a finite set, which represents the amount of the first kind of information.

Actually both meanings are central to the problem of sending messages through a noisy communication channel, which leads to the related and even more abstract measure of *mutual information* between two or more random variables. Shannon's remarkable theorem states that there is a maximum number of messages that can be sent through the channel in a suitably encoded form such that they can be recovered with arbitrarily small error when the messages are long enough. This is true even though the channel introduces errors with a fixed probability of each symbol sent. For good efficiency the messages sent in communication are often very long. For instance, instead of sending short individual messages they are often bundled into long blocks.

In statistics and estimation there are no messages unless by a message we mean a set of observed data, and we certainly are not interested in their number for storing them nor sending them anywhere. What we are interested in are the statistical properties of data, which is the second meaning of information, and its gathering. This process is a virtual synonym of estimation, and it lies at the heart of the theory of estimation in this book. It is possible that the confusion of the two meanings of information is the reason that Shannon's work on communication has never really influenced statistics and estimation, and in fact few statisticians have been studying information theory, the result of which, I think, is the disarray in the present discipline of statistics.

The notion of limits and the asymptotical behavior are not part of the foundation in the theory of estimation in this book. Rather, they are regarded as approximations of the fundamental notions, which can be useful when the real behavior cannot be easily calculated. For instance the Central Limit Theorem (CLT) is useful because the estimates of the most important maximum likelihood (ML) estimator have a distribution which converges to a normal distribution, frequently rapidly, while the exact distribution can be difficult to calculate.

In the rest of this introduction we outline the theory developed here. Although the work is still introductory, we do derive theorems that state fundamental properties of estimators, without which no theory of estimation can be regarded as satisfactory.

In Chapter 2 we discuss the basics of coding, perhaps in greater detail than necessary. The reason is that a case can be made that coding is equivalent to the very idea of probability, which to be sure is abstract, and coding will provide an understanding which is useful both in the fundamental sense and in direct applications. Moreover, the elementary facts are elegant and can be fun to study for anyone. After all, we all do and apply coding in everyday life. Coding is also the basis for the very abstract notion of entropy, the genesis of Shannon's information theory, of which the basics are discussed in Chapter 3. There the second main notion in information theory is the *mutual information*, of which a different new version, the *maximum estimation information*, is introduced together with the *maximum capacity*. These are fundamental in estimation and provide the basis for defining optimality. They also give a wider perspective to the channel coding problem, in which Shannon's channel coding theorem appears as an asymptotic but important special case.

We begin Part II with a general discussion of the estimation problem together with an introduction to the theory in Chapter 4. Just as other theories are based on axioms or, better, postulates so is this theory. There is just one postulate, which as a background requires a class of distributions called *models* of data. The class replaces the naive assumption that the data are generated by a "true" distribution, and the task is to estimate it. Instead of regarding any model to be "true" or "false" we regard it as good, bad, or something in between, which, importantly, can be assessed. This is done by the postulate as the estimation criterion, which in broad terms can be stated thus: the best estimator as a function taking the observed data to a model in a family is one which is determined by the family, and which maximizes the probability of the observed data. But this looks like an impossible task, because no distribution exists under which all data sets have the maximum probability. In fact, for each data set a distribution exists which assigns the probability unity to it. In the past this simple fact has blocked all similar attempts to found estimation theory on probability maximization.

However, there is a way out of the impasse: first, the requirement that the estimator must be determined by the model class prevents favoring any particular data, and, secondly, the maximization of the probability is interpreted as the necessary condition for the maximum an estimator has to satisfy for optimality. For now, let it be that all this can be made precise in a manner which insures optimality of estimators in virtually every sense we can reasonably ask, and nothing is lost by dropping the assumption of the "truth" but a lot is gained with a richer theory as the result.

After the general discussion and the introduction of the postulate criterion three equivalent characterizations of optimal estimators are described, each illuminating different important properties of them. The first is based on the concept of maximum capacity, achieved by the *maximum capacity* (MC) estimator, of which the ML estimator is a special case, and which permits the estimation of the number of parameters, their structure, and even the intervals. The second redefines the same estimators by the necessary conditions for the maximization of the probability on the data. Finally, the third way to define the optimal estimators is by a new *complete* form of the minimum description length (MDL) principle. It sharpens the older *general* MDL principle for model classes, which are narrower but more accurately specified. We also correct the common misunderstanding that the general MDL principle is a special case of Bayesian methods. This is not true, for no priors are needed in the general MDL principle.

In Chapter 5 we prove that the optimal estimators have further desirable properties, similar to the Cramér–Rao inequality but more comprehensive, because they involve the Kullback–Leibler (KL) distance rather than just the covariance. Moreover, the theorems are non-asymptotic, and they include the estimators for both the real-valued parameters and their number. The rest of the chapter is devoted to consistency either in the KL distance or in the sense of almost surely. The former is appropriate for models of "batch" type, fitted to a single finite string, and the latter for models of random process type, needed for prediction. We show that no estimator defined by any criterion exists which beats the MC estimator defined in this book. The chapter ends with a theorem on consistency of the estimates of the number of parameters, which implies that the optimal estimator has the fastest possible convergence rate. This is actually an old theorem on universal coding, whose application to estimation has not been realized due to the lack of an estimation theory.

Chapter 6 is devoted to a new theory of interval estimation, based on the general maximum probability criterion postulate. Because the non-interesting interval consisting of the entire parameter space has the maximum probability unity it is a challenge to construct an estimator that could be called optimal for intervals. However, there is a way, and the result is an estimator of which the usual point-wise ML estimator is the special zero-length interval case. By the CLT an optimally estimated one-dimensional interval has the probability about 0.68, while the optimally estimated k-dimensional interval has the probability 0.68^k. These intervals represent optimal precision on the parameters, which varies from point to point in the parameter space.

Building on these ideas we describe in Chapter 7 the basics of a theory of hypothesis testing, which differs substantially from the existing methods. The hypotheses are either models or sets of models, none of them "true." We consider data to consist of the meaningful information bearing part and two types of "noise," fast fluctuations and slow drift type of variations, which are harder to detect. We model all three of them differently. Hypothesis testing is regarded as the problem of finding out if the observed data are typical for a proposed hypothesis, meaning "acceptance" of the hypothesis, or atypical, "random," relative to the hypothesis, meaning its rejection. The data are represented by test statistics, functions of data, selected so that their distributions are "peaked" in order to have a sharp separation of the typical and the atypical data. In fact, for a number of different types of tests there are test statistics whose distributions are peaked enough to allow acceptance of the hypothesis as sharply as its rejection, which is the only test that can be made in traditional hypothesis testing.

There are two important test statistics and their distributions: the ML estimates with their induced density functions, and the KL distance with a distribution induced by the multinomial. The latter replaces the customary asymptotic χ^2 approximation, which is shown to be grossly inadequate. The same applies to another frequently applied test statistic, the logarithm of the ratio of maximized likelihoods, which asymptotically also admits the χ^2 distribution of appropriate degrees of freedom, but which also is inadequate. It appears that selecting first the χ^2 distribution as done traditionally and then looking for a fitting test statistic is like putting the cart before the horse.

In Chapter 8 we discuss the linear quadratic denoising problem in the MDL framework. We describe both hard and soft thresholding, the latter of which poses a problem to the MDL theory. The novelty over the past publications on linear quadratic denoising is that the MDL criterion has no arbitrary hyperparameters, for they are optimized.

Finally, in Chapter 9 we discuss a different sequential way to estimate models which is appropriate for time series type of random processes. In them the ML estimators are updated not only from the past data, which predictive "plug-in" estimators do, but the updating also includes the most important latest data point. As a result, the so-maximized likelihood is bigger than the batch-wise maximized likelihood. We discuss such models for the class of discrete Markov chains, where the states have variable length as a function of the past data. Their main advantage is that the problem of an exponentially growing number of parameters in the usual m-grams is avoided. We also study linear least squares models, both of the type where the regressor matrix is fixed and of the autoregressive (AR) and autoregressive moving average (ARMA) type, where the matrix is determined by the observations. We derive the recursive predictors in Kalman's theory in a simpler non-redundant manner without the cumbersome Riccati equations. As the main result we give a short proof of the lower bound for the prediction error when the parameters have been estimated.

Although Appendix A includes just the very basic notions of the algorithmic theory of information or complexity, we introduce a new version of complexity in terms of the shortest program, which need not satisfy Chaitin's and Kolmogorov's prefix condition. It was inspired by the ideas in Chapter 4 together with Solomonoff's original idea. We define a non-asymptotic notion of *relative randomness*, which in turn inspired the new theory of hypothesis testing in Chapter 7.

The background knowledge required from the reader is a solid understanding of basic probability theory, however without the need to know the intricate measure theory since the only measurable sets considered are finite and countable sets, as well as open sets and their closures. It is not necessary to know much ordinary statistics, because most of its notions are obsolete and replaced by fewer, different, and more fundamental ones. The basic ideas of information and coding theory which are relevant in estimation are explained in the first part of the book.

Part I

Coding and information

The word "information" has several meanings, the simplest of which has been formalized for communication by Hartley [17] as the logarithm of the number of elements in a finite set. Hence the information in the set $A = \{a, b, c, d\}$ is $\log 4 = 2$ (bits) in the base 2 logarithm. Only two types of logarithm are used in this book: the base 2 logarithm which we write simply as log, and the natural logarithm, written as ln. The amount of information in the set $A = \{a, b, c, d, e, f, g, h\}$ is three bits, and so on. Hence such a formalization has nothing to do with any other meaning of the information communicated, such as the utility or quality. If we were asked to describe one element, say c in the second set, we could do it by saying that it is the third element, which could be done with the binary number 11 in both sets, but the element f in the second set would require three bits, namely 011. So we see that if the number of elements in a set is not a power of 2, we need either the maximum $\lceil \log |A| \rceil$ number of bits or one less, as will be explained in the next section. Hence, we start seeing that "information," relative to a set, could be formalized and measured by the shortest code length with which any element in a set could be described. Again, the word "coding" has nothing to do with description for the purpose of hiding or keeping information secret. It simply means a way to specify the elements in a somehow defined set, which to be sure can be done in a number of ways.

Another interpretation has been suggested for this formalized meaning of information, namely "complexity," for clearly the bigger the set the more complex the description of its elements is in that their coding requires more bits. Big numbers not only require more digits to describe them but also whatever they represent their meaning is hard to comprehend. The classical example is "light year" that physicists use for measuring huge distances that would be hard to comprehend otherwise. By contrast, we understand the size of sets up to five or so without having to count, but the complexity increases surprisingly rapidly with the size of the set. The meaning of the word complexity is not only restricted to the description length of the elements but also to the difficulty of the task of finding an element in a set, which grows with the size of the set.

As we shall see the complexity of the task of estimation can be measured in the same manner, and, in fact, the very idea of probability amounts to the same thing! Whether to use the word "information" or "complexity" depends on which aspect we want to emphasize. There is also a budding theory of complexity of computations, but the difficulties of getting a general theory appear to be orders of magnitude greater, and we regard the word "complexity" as synonymous with the three notions of information, namely, Shannon information, combinatorial information, and algorithmic information, as described in a brief paper by Kolmogorov [21], which are all essentially a code length.

We can rarely formalize intuitive ideas entirely, because the complexity of the formal description grows rapidly beyond our means. We only need to bear in mind the language used in law books, which tries to cover all the contingencies of the law and which is nearly indecipherable. The formal definition of information covers only the aspect of the word that relates to the code length. Hence it says nothing about whether the information conveyed is useful for anything or is just "noise." However, the code length turns out to be equivalent to probability and hence adds to our understanding of this fundamental but elusive idea. Moreover, the objective in estimation and statistics is to extract "information" from data, and rather than settle for the intuitive ideas it seems to be of interest to see if the code length interpretation can after all suggest a sense of the quality of the "information" by separating the useless noise from the useful part.

2 Basics of coding

We begin with the description of the most basic and primitive codes. Let $A = a_1, \ldots, a_k$ be a finite set: the set is called an *alphabet*, and its elements are called *symbols*. The main interest is in the set of finite sequences $s^n = a_i a_j \ldots$ of the symbols of some length n, called *messages* or for us just *data*. The problem is to send or store them in a manner that costs the sending device little time and storage space. Again for practical reasons these devices use binary symbols 0 and 1, while the original symbols are represented as a sequence of binary symbols, such as the eight bits long "bytes." Let for each symbol a in A, $C : a \mapsto C(a)$ be a one-to-one map, called a *code*, from the alphabet into the set of binary strings. It is extended to the messages by *concatenation* $C : a_i a_j \ldots a_n \mapsto C(a_i)C(a_j)\ldots C(a_n)$. Both binary strings $C(a)$ and $C(a_i)C(a_j)\ldots C(a_n)$ are called *codewords*.

This will give us an upside down binary tree, with the root on top, whose nodes are the codewords, first of the symbols for $n = 1$ and then of the length 2 messages, and so on. The left hand tree in Figure 2.1 illustrates the code for the symbols of the alphabet $A = \{a, b, c\}$. The extension of the codes from symbols to sequences by concatenation creates a problem: We would like to decode the message string symbol for symbol from the binary string representing the codeword of the message. This can be done if we put the restriction on the code trees that only the leaves can be codewords. This means that no prefix of a codeword can be a codeword of another symbol. The right hand tree in Figure 2.1 illustrates such a *prefix* code. This permits decoding of the strings without commas to separate the codewords for the symbols by just climbing down the code tree starting at the root until a leaf is met. As an example, unlike with the left hand tree, the codeword $C(0110001)$ defined by the right hand tree decodes as the string *bcab*. A code is called *complete* if all the leaves are codewords.

12 Basics of coding

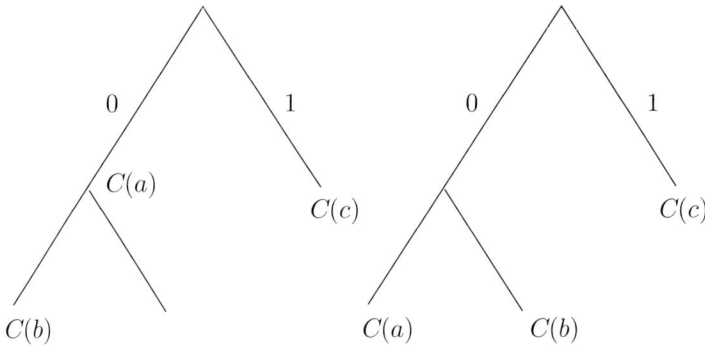

an incomplete non-prefix code   a complete prefix code

Figure 2.1 Two codes for the alphabet $\{a, b, c\}$.

It is not always necessary to construct the code trees, which for large alphabets would be awkward. Nor is it necessary that the domain of a concatenation map consists of all finite messages A^*, because we only need to encode the "meaningful" messages, where again the meaning does not refer to the content but to the syntax or grammatical correctness. When these can be defined by an algorithm efficient coding is quite easy. We could sort the meaningful messages first by the length n and then lexically after having sorted the alphabet. We encode each string s^n of length n as the smallest binary integer larger than all the ordinals of the strings smaller than s^n. Its length is a bit longer than the binary logarithm of the integer; see Appendix B. As a matter of fact, messages written in a natural language often have a grammar, which at least approximately permits us to decide which messages are well formed. The natural languages tend to be too complex to admit an algorithm for their syntax, and we cannot tell exactly which sentences are well formed, but having even a partial grammar reduces the domain of the map and hence the image.

2.1 Coding and probability

The lengths of the prefix codewords satisfy the fundamental inequality due to Kraft:

$$\sum_{a \in A} 2^{-|C(a)|} \leq 1, \tag{2.1}$$

where $|x|$ denotes the length of the binary string x. Moreover, the equality holds if and only if the code is complete. The proof is immediate: If you represent the length zero codeword by the root, the equality holds. It also holds for length 1 codewords. By applying the inequality

$$2^{-|s0|} + 2^{-|s1|} \leq 2^{-|s|} \tag{2.2}$$

recursively starting at the root we see that the inequality is retained by summing over all leaves $C(a)$. The equality holds if all the leaves are codewords, which is the case in a complete tree.

Conversely, for any set of natural numbers $n_1, \ldots, n_m$ satisfying

$$\sum_i 2^{-n_i} \leq 1 \tag{2.3}$$

a prefix code exists such that $|C(i)| = n_i$. To see this sort the lengths $n_1 \leq n_2 \leq \cdots \leq n_m$. Starting at the root mark the left most path $00\ldots0$ of length n_1 as the first codeword. Continue in lexical order, $s0 < s1$, by taking the left most path of those remaining that have length n_2, excluding the extension of the previous path, as the next codeword. Continue until all codewords have been constructed as leaves. As an example, consider the numbers $\{n_i\} = \{2,2,2,3,3\}$. We get the complete prefix code $C = \{00, 01, 10, 110, 111\}$.

What this means is that a complete prefix code, or rather the code lengths, satisfies the Kraft inequality with equality and defines a distribution $\{2^{-n_i}\}$ for A, and, conversely, given a distribution $P(a)$ on A, the integers $n_a = \lceil \log 1/P(a) \rceil$ define a prefix code – albeit not necessarily a complete one. Since we do not need repeated data for this to hold and since we can design prefix codes for all finite and even countable sets we have a common concrete interpretation of probability, which is applicable to elements of sets, regardless of whether the elements repeat or not. In particular, we do not need the awkward attempts to redefine probabilities as "degrees of belief" in the Bayesian philosophy. Disregard the degree of belief, and just construct a prefix tree for the elements of interest using logic, mathematics, or whatever you wish, and the probabilities are defined. Since the resulting probabilities satisfy Kolmogorov's axioms there is no need to interpret them as degrees of belief. Moreover, since the definition of the probabilities is based on logic we can justify our findings and even talk about optimizing them, which of course is a landmark of scientific procedures. Clearly, the Kraft inequality generalizes to binary trees with countable leaves, because it holds for all finite subsets, and by limiting processes we can further extend this interpretation of probability to open sets and their closures, which are all the measurable sets we need to consider in this book and all statistics for that matter. It is clear that for an infinite countable alphabet we need an algorithm to encode any element of the alphabet in a prefix manner such that the Kraft inequality holds.

There can be several complete prefix trees for the same set A, and it will be instructive to discuss one of them. Let m be the integer which satisfies $2^{m-1} < |A| = n \leq 2^m$, and put $r = 2^m - n$. Encode r elements of A, say the last r in some enumeration of A, with the code length $m-1$, and the rest with length m. Call such a tree almost *balanced*, while the term balanced tree is reserved for one of 2^m leaves for some m. We verify the Kraft equality: $(n-r)2^{-m} + r2^{-(m-1)} = r2^{-m} + n2^{-m} = 1$. Any reduction of the maximum code length from m will make the Kraft sum greater than unity and hence the code would not satisfy the prefix condition.

Without the prefix requirement one can argue that at most one half of the 2^m strings of length m can be encoded with a codeword length less than $m-1$, one quarter with a codeword length less than $m-2$, and so on, which implies that for large sets only the fraction 2^{k-m} of all the elements can be encoded with less than k bits from the maximum m. This follows from the fact that there are not very many short strings: $\sum_{i=1}^{k} < 2^{k+1}$. By adding the complete prefix requirement we can strengthen this: if we interpret binary strings as codewords, and since it is always possible to encode members of a set by their ordinals up to the maximum, then the leaves of the almost balanced tree are in a sense the most efficient way to encode the elements of the set with a prefix code. This is clearly what we would like to capture by the notion of *complexity* of the set A, which should be $\lceil \log |A| \rceil$ rather than Hartley's early and insightful definition of information $\log |A|$, which, as we shall see, Shannon generalized.

Another extreme complete prefix code is obtained by encoding one element, say a_1 with one bit, the next with two bits, and so on, the last two with $n-1$ bits. Again the Kraft equality holds: $2^{-1} + 2^{-2} + \cdots + 2 \times 2^{-(n-1)} = 1$. This tree is maximally "skewed" in that the maximum code length, $n-1$, is also maximum among all complete prefix trees with n leaves.

2.2 Coding and entropy

A giant step in coding, in fact one which forms the basis for the entire information theory, was taken by Shannon. Consider the n-fold Cartesian product A^n of an alphabet A. If we have a distribution $P = \{p_1, \ldots, p_k\}$ defined on the elements of A, $p_i = P(a_i)$, we can extend it to A^n by independence: $P(a_i a_j \ldots) = p_i p_j \ldots$. We feel intuitively that the distribution restricts the complete freedom of association allowed by the Cartesian product, which reduces the size $|A|^n$, where $|A|$ is the size of A. Consider the product of the inverse powers

$$V_P(A) = \prod_i p_i^{-p_i} \qquad (2.4)$$

as a kind of "relevant" size of A and A^n, and take the logarithm $\log V_P(A) = \sum_i p_i \log 1/p_i = H(P)$, the Shannon *entropy*, as the measure of the amount of the *mean information* in A, if we regard $\log 1/p_i$ as the *information* in the element a_i. It follows at once by the independence that

$$\log V_P(A^n) = n H(P). \qquad (2.5)$$

The formal definition of the intuitive and abstract idea of the "relevant size" of A, (2.4), is given by the following theorem, called the noise-free coding theorem, which is a truly fundamental result in all information theory.

Theorem 2.1 *(McMillan and Doob) For all prefix codes for A with distribution P, each defined by the lengths $\{n_i\}$, the mean code length $L(P)$ satisfies*

$$(i) \quad \sum_i p_i n_i \geq \sum_i p_i \log 1/p_i = H(P), \qquad (2.6)$$

$$(ii) \quad L(P) = H(P) \Leftrightarrow p_i = 2^{-n_i}, \text{ all } i, \qquad (2.7)$$

where $0 \log 0 = 0$.

This theorem gives a concrete interpretation of the entropy in terms of the mean code length, and there is no need to talk about the entropy as a "measure of uncertainty" or "disorder," as is often done by well-meaning authors, which just replaces one abstract notion by other even more abstract and undefined notions.

Proof We have

$$\sum p_i \log \frac{2^{-n_i}}{p_i} = (\log e) \sum p_i \ln \frac{2^{-n_i}}{p_i} \tag{2.8}$$

$$\leq (\log e) \sum p_i \left[\frac{2^{-n_i}}{p_i} - 1 \right] \tag{2.9}$$

$$= (\log e) \left[\sum 2^{-n_i} - 1 \right] \leq 0. \tag{2.10}$$

In the first inequality we used the fact that $\ln x \leq x - 1$.

The same proof applies even when the probabilities p_i and 2^{-n_i} are replaced by densities $p(x)$ and $q(x)$, and we get the important KL distance between two density functions $p = p(\cdot)$ and $q = q(\cdot)$

$$D(p\|q) = \int p(x) \log \frac{p(x)}{q(x)} dx \geq 0. \tag{2.11}$$

Notice, however, that $D(p\|q)$ is not symmetric nor does it satisfy the triangle inequality, so that it is not a metric. If we regard $\log 1/p(x)$ as an ideal code length, because it is not integer-valued unlike real code lengths, we see that the KL distance is a mean (ideal) code length difference. It is also called the *cross entropy*. We make repeated use of this distance measure. This is a good example of the phenomenon that a trivial corollary of a fundamental code fact can have highly non-obvious applications beyond coding.

2.3 Huffman's code

The next important step in coding theory is to find a prefix code C for a set A with a probability distribution $P(a_i)$, defined on the symbols, which minimizes the mean code length:

$$L(C) = \sum_i P(a_i)|C(a_i)| = \sum_i p_i n_i. \tag{2.12}$$

The minimum mean length prefix code can be found by the neat Huffman algorithm [18], which we describe mostly because of its elegance and historical value; it no longer plays the dominant role in applications of coding theory it used to play. The algorithm starts with the elements of $A = \{a_1, a_2, \ldots, a_m\}$ sorted by the decreasing probabilities $p_1 \geq p_2 \geq \cdots \geq p_m$. The order of elements with equal probabilities does not matter. We shall derive the characteristic properties of an assumed optimal code $C = C_m$ for A which determine it. If we compare the assumed minimum $p_i n_i + p_{i+1} n_{i+1}$ with $p_i n_{i+1} + p_{i+1} n_i$ we see that the difference is $(n_i - n_{i+1})(p_i - p_{i+1})$, which for minimality should be equal to or less than zero, because switching the lengths from the minimum must not decrease the sum $p_i n_i + p_{i+1} n_{i+1}$. For this to be the case $n_i \leq n_{i+1}$.

Also the optimal code must be complete. This implies that the sibling of a deepest leaf of length n_m also has the same length: $n_{m-1} = n_m$. Their mother node has the probability $p_{m-1} + p_m$ and length $n_m - 1$. We now have an alphabet of size $m - 1$ with a distribution, where the mother node is a new leaf replacing the two siblings, while the other $m - 2$ leaves are those of C_m with the same probabilities. Clearly, the new subtree C_{m-1} with $m - 1$ leaves must be optimal, too. Hence by sorting the new set of $m - 1$ symbols by decreasing probabilities and repeating the arguments above we eventually arrive at the two-leaf subtree, which necessarily is optimal.

If we encode strings of A^n by the n-fold concatenation of the leaves of a Huffman tree for A we get a complete prefix tree, which, however, is not optimal. This is obvious if we take $A = \{0, 1\}$, for which the Huffman tree must be the two-leaf tree no matter what the probability $P(0)$ is, and $L_P(A) = 1$. Hence, $L_P(A^n) = n$ for all n, which also is the maximum possible entropy no matter what the distribution on A^n is. Hence, in order to have a prefix tree close to the entropy we must calculate the distribution for some extension of A, say A^d, and construct an optimal tree for this with the induced distribution P^d. The selection of d depends on the size $|A|$ and how close to the per symbol entropy $H(P^d)/d - h(P(0))$ we want to be. It is customary to denote the binary entropy by the lower case letter $h(p)$, where $P(0)$ or $P(1)$ is denoted by p.

Denoting the extended d-fold alphabet by A for simplicity, the mean code length of the further extension A^n is given by

$$L_P(A^n) = \sum_{s \in A^n} P(s) L(s) = n L_P(A), \tag{2.13}$$

where $L(s) = n_i + n_j + \cdots$ for $s = a_i, a_j, \ldots$. We prove this in the example for the alphabet $A = \{a, b, c\}$ and the prefix code in Figure 2.1 with $P(a) = p_a$, $P(b) = p_b$, and $P(c) = p_c$. For the sequence $s = bca$ we have $P(s) = p_b p_c p_a$, $L(s) = n_b + n_c + n_a = 5$, and $L_P(A) = 2p_b + 2p_a + p_c$. If we extend each leaf by the leaves of the tree C, the new leaves are two bits longer from nodes $C(b)$ and

$C(a)$ and one bit longer from the node $C(c)$, with the result that the mean length of the new tree is $L_P(A^2) = 2L_P(A)$. We see that the mean code length shares the property of the entropy for the extension by independence: $H(P^n) = nH(P)$.

2.4 Codes as number representations

One trouble with Huffman's codes is that the whole tree is required before we can encode any element, which makes recursive coding difficult for strings of symbols. Another even greater problem, shared by all concatenation codes, is the need for alphabet extension. Because both of the binary symbols require one bit, the mean code length for the binary alphabet for all distributions is 1. As seen above, in order to get compression for binary strings we must extend the alphabet to blocks of, say, d symbols, and design a Huffman code for that. Although this can be done not only for Bernoulli models but even for binary strings modeled by Markov chains, the book-keeping of the separate extensions needed at each state becomes elaborate and awkward. These difficulties can be avoided by totally different types of coding, based on number representations, of which *arithmetic* coding [36] is currently the dominant form of coding. In such codes no code tree is needed because both the code and the coding operations are described by an algorithm. This is particularly important in the so-called universal coding in which the codes are designed for distributions estimated from the data. Although this clearly suggests an intimate link between coding and estimation, the two problems are not identical and should not be confused. In coding there is no need for explicit estimation of parameters, which obviously is the requirement in estimation.

What attracted me to this type of coding was the original idea of probability based on combinatorics, which amounts to counting the number of things subject to certain restrictions. Unlike in counting and creating probabilities by counting, where only formulas are required, we need algorithms to define codes for practical reasons. It seems to be of some interest to reproduce here a part of a paper on coding based on multiple radices I wrote in 1979 [38] in an attempt to place arithmetic codes in a proper mathematical basis, after having introduced arithmetic coding [36] in a more or less intuitive manner. The multiple radix representations can be found in text books on combinatorics, without any connection to coding or probabilities.

We start with the familiar single radix representation of binary sequences as numbers and the associated reverse function. The codewords then are ordered numbers. Let $s = 01010010$ be a message. If the jth 1 of s is the $i^*(j)$th symbol, counted from the right hand end, we may associate with s the number as its code,

$$C^*(s) = \sum_{j=1}^{m(s)} q^{i^*(j)-1}, \qquad (2.14)$$

$$C^*(s) = 0, \text{ if } m(s) = 0, \qquad (2.15)$$

where q is a positive number, called the *radix*. In the example $i^*(1) = 2$, $i^*(2) = 5$, and $i^*(3) = 7$. The representation is one-to-one and hence decodable, except for the left-most zeros in s, if and only if $q \geq 2$. This follows from the condition that each power in the sum is larger than the sum of the smaller powers. With this property the recovery of s from $C^*(s)$ proceeds as follows: The position $i^*(m(s))$ of the left-most 1 of s is the smallest power of q, but greater than $C^*(s)$, and after $q^{i^*(m(s))-1}$ is subtracted from $C^*(s)$, the cycle is repeated.

Dually, let $i(j)$ denote the position of the jth 1 in s, both counted from the left hand end. This gives the second representation of s:

$$C(s) = \sum_{j=1}^{m(s)} q^{-i(j)}, \qquad (2.16)$$

$$C(s) = 0, \text{ if } m(s) = 0. \qquad (2.17)$$

For $q \geq 2$ this, too, is one-to-one except for the right-most zeros in s. The decoding process is analogous to the previous case. Observe, however, that with (2.14) the last encoded 1 is decoded first, last-in-first-out (LIFO), while with (2.16) the first encoded 1 is decoded first, first-in-first-out (FIFO), which is important in communication: the symbols can be decoded just as soon as they have been encoded and sent to the decoder. In the LIFO version the entire message must be sent before it can be decoded.

Finally, $C^*(s)$ is the ordinal of the sequence s in the set of all sequences without the leading zeros 1, 10, 11, 100, . . ., while $C(s)$ is the normalized ordinal as the cumulative probability, obtained by dividing $C^*(s)$ by the least power of q which dominates $\max_{\bar{s}:|\bar{s}|=|s|} C^*(\bar{s})$, where $|s|$ is the length of s. A similar interpretation pervades the generalizations to follow.

Now let q_0 and q_1 be two *radices*, one for each of the two symbols 0 and 1, respectively, with both numbers greater than 1. Using the notation of (2.14) put

$$C^*(s) = \sum_{j=1}^{m(s)} q_0^{i^*(j)-j-1} \times q_1^j, \tag{2.18}$$

$$C^*(s) = 0, \text{ if } m(s) = 0. \tag{2.19}$$

To simplify matters we consider the case where both the encoder and the decoder know the number $m = m(s)$. As in the single radix case we want to select the radices so that each term $q_0^{i^*(j)-j-1}$ in the sum $C^*(s)$ is greater than the sum of the smaller terms. This allows us to decode in all strings the left-most 1 by finding the $i^*(m)$ as the smallest n such that the single term $q_0^{n-m} q_1^m$ exceeds the maximum possible value of the sum $C^*(s)$. We now find the conditions of q_0 and q_1 when the decoding works.

Because $q_0 > 1$ the maximum $C^*(s)$ for strings having m 1s is obtained for the string $\hat{s} = 1\ldots 10\ldots 0$, which maximizes the indices $i^*(j)$ and hence the sum (2.18). Therefore, in order for the decoder to find $i^*(m)$, written as n, correctly it must be the case that

$$q_0^{n-m} q_1^m > C^*(\hat{s}) = q_0^{n-m-1}(q_1^m + \cdots + q_1) = q_0^{n-m-1} q_1 \frac{q_1^m - 1}{q_1 - 1}. \tag{2.20}$$

Equivalently $q_1^{1-m} + q_0 q_1 > q_0 + q_1$, which is true for every m if and only if

$$1/q_0 + 1/q_1 \equiv 2^{-\log q_0} + 2^{-\log q_1} \leq 1. \tag{2.21}$$

In the single radix case $q_0 = q_1 = q$ this reduces to the earlier stated condition: $q \geq 2$ for decodability. If we put $p_0 = 1/q_0$ and $p_1 = 1/q_1$, we get the condition for decodability as a generalized Kraft inequality, where the role of the integer-valued code lengths is played by the real numbers $\log 1/p_i$ enforcing the intimate link between code lengths and probabilities.

We decode the string above $s = 01010010$, for which $i^*(1) = 2$, $i^*(2) = 5$, and $i^*(3) = 7$, and the sum (2.18) is given by $C^*(s) = q_1 + q_0^2 q_1^2 + q_0^3 q_1^3$. The smallest n such that $q_0^{n-3} q_1^3 > C^*(s)$ is $n = 7$. Subtracting the maximum term $q_0^{7-3-1} q_1^3$ from $C^*(s)$ gives the remaining code, and the next position of 1 for $n = 5$, and so on until the last position of 1 for $n = 2$.

The interesting and important feature of the representation (2.18) for coding is that for strings with many more 0s than 1s the number C^* grows nowhere near as $2^{i^*(m)}$, if we pick q_0 only slightly greater than 1 and q_1 as small as condition (2.21) permits, which gives compression unlike in the single radix case.

We next derive the dual code as an introduction to the final arithmetic codes. Divide the terms in the sum (2.18) by the left hand side of (2.20), i.e. the minimum term that dominates $C^*(s)$:

$$C(s) = \sum_{j=1}^{m(s)} q_0^{i^*(j)-n+m-j-1} \times q_1^{j-m}, \qquad (2.22)$$

$$C(s) = 0, \text{ if } m(s) = 0. \qquad (2.23)$$

By changing the indexing to run from right to left, i.e. by replacing j by $m - j + 1$ and i^* by $i = n - i^* + 1$, we get

$$C(s) = \sum_{j=1}^{m} p_0^{i(j)-j+1} \times p_1^{j-1}, \; C(s) = 0 \text{ if } m = 0. \qquad (2.24)$$

The right-most 0s do not change the value of $C(s)$.

2.4.1 Arithmetic codes

Although the basic idea of arithmetic codes is the dual number representation (2.24) a more convenient way to derive the needed recursions is to recognize it as the cumulative probability of alphabetically ordered strings of each length n, where the two radices are defined by the conditional probabilities $p(0|x^n)$, $p(1|x^n) = 1 - p(0|x^n)$, including the empty string $x^0 = \lambda$, for which $p(0|\lambda) = p(0)$. The probability of string x^k is given by

$$p(x^k) = \prod_{i=1}^{k} p(x_i | x^{(i-1)}). \qquad (2.25)$$

It was known to Shannon that since the cumulative probabilities $C(x^k) = \sum_{y^k < x^k} p(y^k)$ define a monotonically increasing function of the lexically ordered strings they could be used as codewords for the strings. However, the coding theory at the time, maybe because of Huffman's optimal codes, was developed using concatenation codes. The feeling was that there is no need to improve an optimal code, and this idea lay dormant until the introduction of arithmetic coding in 1976 [36]. As a final comment, because of patents on arithmetic codes by IBM

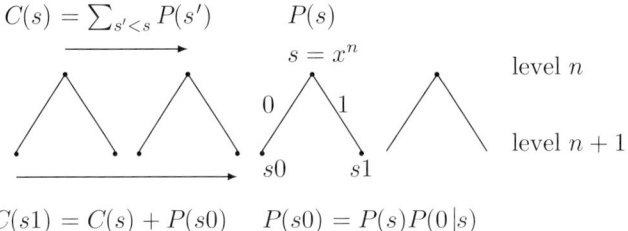

Figure 2.2 A binary arithmetic code.

they were not widely known nor used until the 1990s when the patent protection expired, and their use exploded. We give in Figure 2.2 the simple derivation of the basic recursions

$$p(x^{i+1}) = p(x^i)p(x_{i+1}|x^i), \tag{2.26}$$
$$C(x^i 0) = C(x^i), \tag{2.27}$$
$$C(x^i 1) = C(x^i) + p(x^i 0), \ i \geq 1, \tag{2.28}$$

where we have written $x^i 0$ for the string of length $i+1$ ending in 0. These equations generalize easily to non-binary alphabets.

The problem that a codeword as a binary fractional number increases only when the string ends in 1 can be overcome by representing the binary numbers as binary strings. A more serious problem is the precision problem: an ever increasing register is needed for the product on the first line, and with a fixed size register the problem arises of what to do with the addition on the third line when the register is full. The solution of the first problem turns out to be easy. We reduce the probabilities to a fixed precision, say to q significant bits, thus

$$\bar{p}(x^{i+1}) = \lfloor \bar{p}(x^i)\bar{p}(x_{i+1}|x^i) \rfloor_q. \tag{2.29}$$

The loss in the mean code length is $O(2^{-q})$ per symbol, which can be reduced by increasing the register size just as the per symbol loss in concatenation codes can be reduced by increasing the block length. Since $p(x^i)$ decreases while the binary number $C(x^i)$ grows with a growing string, the addition on the third line generally affects only the last q bits of the fractional number $C(x^i)$, except when overflow occasionally propagates left. This can be handled by a "bit stuffing" technique with the register of size $q + 1$, where the first bit position is normally 0 but is switched to 1 by the encoder when the register becomes full signaling the event to the decoder which can perform the correct subtraction needed. As a final comment, any concatenation code can be implemented as an arithmetic code for non-binary alphabets with the fixed conditional probabilities given by the number 2 raised to the negative powers of the code words. An example is given in [45], page 38.

We see that no code tree need be constructed. Rather, the encoder needs just the probability $\bar{p}(x_{i+1}|x^i)$ modeled for each symbol to encode the symbol, while the decoder needs only the past coded string and the same probability to recover the symbol. The result is that the critical problem in data compression is how to model the data. This led to the general minimum description length (MDL) principle for estimation, which, in turn, will be sharpened to the complete MDL principle in this book.

3 Basics of information

3.1 Main definitions

We give the basic properties of the entropy, all of which follow from the noiseless coding theorem. We use the following notation: when a random variable (r.v.) X has the distribution P, which is symbolized as $X \sim P$, the entropy $H(P)$ is often written as $H(X)$. Clearly,

$$H(X) \geq 0, \tag{3.1}$$

because it is the sum of non-negative elements. In the important binary case, where $P = \{p, 1-p\}$, the entropy is often written as $h(p)$. It is a symmetric concave function of p, reaching its maximum value 1 at $p = 1/2$, and vanishing at points 0 and 1. If the joint of X and Y is given by

$$(X, Y) \sim p(x, y), \tag{3.2}$$

then the *conditional* entropy is defined as

$$H(Y|X) = \sum_{x,y} p(x, y) \log \frac{1}{p(y|x)} \tag{3.3}$$

$$= \sum_x p(x) \sum_y p(y|x) \log \frac{1}{p(y|x)} \tag{3.4}$$

$$= \sum_x p(x) H(Y|X=x). \tag{3.5}$$

Notice that $H(X|X) = 0$, and $H(Y|X) = H(Y)$ if X and Y are independent:

$$H(X, Y) = \sum_{x,y} p(x, y) \log 1/p(x, y) \tag{3.6}$$

$$= \sum_x p(x) \sum_y p(y|x) [\log 1/p(x) + \log 1/p(y|x)] \tag{3.7}$$

$$= H(X) + H(Y|X). \tag{3.8}$$

The entropy has a distinctive grouping property. If we lump together a part of the symbols, say in a subset $S = \{a_1, \ldots, a_k\}$ of the alphabet $A = \{a_1, \ldots, a_n\}$, which has the distribution $p = \{p(a_1), \ldots, p(a_n)\}$, then

$$H(p) = h(P_S) + P_S H(\{p(a_1)/P_S, \ldots, p(a_k)/P_S\})$$
$$+ (1 - P_S)H(\{p(a_{k+1})/(1 - P_S), \ldots, p(a_n)/(1 - P_S)\}), \quad (3.9)$$

where $P_S = \sum_{i=1}^{k} p(a_i)$. Hence, the entropy remains the same when we add to the entropy of the binary event S the entropies of the conditional events defined by the outcomes within S and A/S. The interested reader may want to verify this.

The important *mutual information* deserves its own section.

3.2 Mutual information

3.2.1 Expected mutual information

We begin with an outline of Shannon's (expected) mutual information and its maximum, the *channel capacity*. Consider a joint random variable X, Y with a distribution $p(x,y)$, which we can write as $p(x,y) = p(x)p(y|x)$, where $p(x) = \sum_y p(x,y)$ and similarly for $p(y)$. The variables X and Y are assumed to range over finite sets. One way to measure the strength of the link between X and Y is by the joint entropy $H(X,Y) = H(X) + H(Y|X) = H(Y,X)$. However, this can be large or small simply because the entropy $H(Y)$ or $H(X)$ is large or small, which does not capture what we want. Shannon considered the mean of the ideal code length difference $\log 1/p(y) - 1/\log(y|x)$ a better measure, and he defined the KL distance between the joint and the product of the margins

$$I(Y;X) = D(p(X,Y)\|p(X)p(Y)) = \sum_{x,y} p(x,y) \log \frac{p(x,y)}{p(x)p(y)} \quad (3.10)$$

$$= \sum_{y,x} p(x,y) \log \frac{p(y|x)}{p(y)} \quad (3.11)$$

to be the (mean or expected) *mutual information* between Y and X. It is non-negative, and zero if and only if X and Y are independent. Because $p(x,y) = p(y,x)$ the mutual information is symmetric: $I(X;Y) = I(Y;X)$.

Further, $I(X;Y) = H(X) - H(X|Y) = H(Y) - H(Y|X)$. To see this, write

$$I(X;Y) = \sum_{x,y} p(x,y) \log \frac{p(x|y)q(y)}{p(x)q(y)} \qquad (3.12)$$

$$= \sum_{x,y} p(x,y) \log \frac{1}{p(x)} - \sum_{x,y} p(x,y) \log \frac{1}{p(x|y)} \qquad (3.13)$$

$$= \sum_{x} p(x) \log \frac{1}{p(x)} - \sum_{x} p(x) \sum_{y} p(x|y) \log \frac{1}{p(x|y)} \qquad (3.14)$$

$$= H(X) - H(X|Y). \qquad (3.15)$$

To add more to intuition, if the mutual information is small, the values of X and Y tell little about the values of each other, i.e. X and Y are nearly independent. Conversely, if it is large, then the joint is far from independent, and the values of X and Y help a lot in coding the values of each other. So it is the difference between the entropy of Y or X and the conditional entropy $H(Y|X)$ or $H(X|Y)$, respectively, that counts.

In communication the objective is to send input sequences x^n, one symbol at a time, through a channel, which introduces errors modeled by a set of conditional distributions, $p(y|x)$, one for each x, where we take for simplicity the "input" x and the "output" y to range over a common alphabet, say the binary. An example is the binary symmetric channel, defined by $p(y|0)$: $p(0|0) = p$, $p(1|0) = q = 1 - p$, and $p(y|1)$: $p(1|1) = p$, $p(0|1) = q = 1 - p$. This is generalized to sequences by independence: $p(y^n|x^n) = \prod_1^n p(y_i|x_i)$. If we have $p > q$ and decode using the identity function $\hat{x}^n(y^n) = y^n$, the probability of error per symbol is q when the binary symbols are sent through the channel, and there is not much to do about the error probabilities if the messages to be sent are allowed to consist of all the possible 2^n input sequences. The problem becomes interesting if we select only a subset of the input sequences as messages to be sent. For instance, if we send just two input sequences, $0^n = 0, 0, \ldots, 0$ and 1^n, we could recover them with small error probability if we decode using the majority rule, i.e. decode sequence y^n as $\bar{x}^n = 0^n = 0, 0, \ldots, 0$ if the number of zeros in y^n exceeds the number of 1's, and $\bar{x}^n = 1^n = 1, 1, \ldots, 1$ in the opposite case, while picking the all zero case if the numbers are equal. Shannon considered the question of the maximum number of input sequences that could be sent through the channel such that the per symbol error probability does not exceed ϵ, for all ϵ, as the length n of the sequences grows to infinity.

In fact, he also considered non-symmetric channels, where the distribution of the "output" sequences y^n can be altered by the choice of the input distribution $w(x^n)$, which affects the error probabilities. With this one converts the family $\{p(y^n|x^n) : x^n \in \{0,1\}^n\}$ of 2^n distributions and the same number of random variables to just two X^n and Y^n with the joint distribution $p(y^n, x^n) = p(y^n|x^n)w(x^n)$. Shannon's great insight was to pick $w(x^n) = w^*(x^n)$ so as to maximize the (mean) mutual information

$$\max_w I_w(Y^n; X^n) = \sum_{x^n} w^*(x^n) \sum_{y^n} p(y^n|x^n) \log \frac{p(y^n|x^n)}{\sum_{x^n} p(y^n|x^n)w^*(x^n)}$$
$$= H(Y^n) - H(Y^n|X^n). \qquad (3.16)$$

The maximized result is called the *channel capacity*. We discuss briefly the reason for this name in the next subsection.

3.2.2 Maximum estimation information

Consider a family of distributions $\{p(y;x) : y \in A, \ x \in B\}$, one for each x, with both sets A and B finite. The role of x is that of a parameter. Later, in estimation the ranges are infinite, in which case the sums are replaced by integrals. Let $\bar{x} : y \mapsto \bar{x}(y) \in B$ denote an estimator function. It defines two random variables Y and $\bar{X} = \bar{X}(Y)$, with the joint distribution

$$\bar{p}(y) = \frac{p(y; \bar{x}(y))}{\bar{C}} = p(y|\bar{x}(y))p(\bar{x}(y)), \qquad (3.17)$$

$$\bar{C} = \sum_y p(y; \bar{x}(y)), \qquad (3.18)$$

$$p(y|\bar{x}(y)) = p(y; \bar{x}(y))/w(\bar{x}(y)), \qquad (3.19)$$

$$p(\bar{x}) = w(\bar{x})/\bar{C}, \qquad (3.20)$$

where $w(\bar{x}) = \sum_{u:\bar{x}(u)=\bar{x}} p(u;\bar{x})$. Notice in particular that the range of y of each $p(y;x)$ is the full range A, while that of $p(y|\bar{x}(y))$ is a proper subset of A, and the two distributions are in general very different.

Define the *maximum capacity* as

$$\max_{\bar{x}(\cdot)} \log \bar{C} = \log \hat{C} \geq 0, \qquad (3.21)$$

the maximum reached with the maximum likelihood estimator $\hat{x}(y)$: $\max_{x \in B} p(y; x)$. It is non-negative, because $p(y; \hat{x}(y)) \geq p(y; x)$ for all x and $\sum_y p(y; x) = 1$. This turns out to be the main concept on which this theory of estimation is founded. We can give it an interpretation as a measure of the maximum amount of information an estimator can obtain about the family of the distributions. Call

$$\max_{\bar{x}(\cdot)} \log \frac{p(y; \bar{x}(y))}{\bar{p}(y)} = \log \frac{p(y; \hat{x}(y))}{\hat{p}(y)} = \log \hat{C} \qquad (3.22)$$

the *maximum estimation information*. We see that this mimics the ratio of the joint and the product of the marginals in (3.16) in that $p(y; \hat{x}(y))$ in the numerator plays the role of the conditional, although it fails to be a conditional, because $\sum_{u:\hat{x}(u)=\hat{x}} p(u; \hat{x})$ is not unity. However, because of this, unlike in the mutual information where the expectation is needed to keep it non-negative, no expectation is needed here. We give ample evidence in Part II that this is the right measure of the desired information rather than the quite different mutual information $\hat{I}(Y; \hat{X})$ in Shannon's sense between the random variables Y with the distribution $\hat{p}(y)$ and $\hat{X}$ with the distribution $p(\hat{x})$. In fact, $\hat{p}(y)$ equals the joint $\hat{p}(y, \hat{x}(y)) = \hat{p}(y|\hat{x}(y))p(\hat{x}(y))$ and the mean of the logarithm of the joint divided by the product of the marginals gives

$$I(Y; \hat{X}) = \sum_{y, \hat{x}(y)} \hat{p}(y, \hat{x}(y)) \log \frac{\hat{p}(y)}{\hat{p}(y)p(\hat{x}(y))} = H(\hat{X}). \qquad (3.23)$$

Although this is not the maximum of the mutual information $\max_{\bar{x}(\cdot)} H(\bar{X})$ of Y with the distribution $\bar{p}(y)$ and $\bar{X}$ with the distribution $\bar{x}(y)$, even the mutual information $H(\hat{X})$ plays a role in estimation. It measures the size of the class of the models $\{p(y; x) : x \in B\}$ in terms of the entropy, and it reflects the mean complexity of the estimators. In so doing there are special cases where the ML estimator does make $p(\hat{x})$ uniform. An example is given by the estimation of intervals in Chapter 6.

Example We give a simple example of three distributions $p(y; x)$ for the "parameters" $x \in B = \{1, 2, 3\}$ and y ranging over $A = \{1, 2, 3, 4\}$. The three distributions are given in the columns of the 4×3 matrix, and the maximizing parameter $\hat{x}$ for each y is indicated in each row. For instance, if $p(4; 2)$ is greater than both $p(4; 1)$ and $p(4; 3)$, then we write $\hat{2}$. For the other rows, in which the ML estimates

3.2 Mutual information

are assumed to be 2, 1, and 3, respectively:

$$p(4;1)\ p(4;\hat{2})\ p(4;3) \\ p(3;1)\ p(3;\hat{2})\ p(3;3) \\ p(2;\hat{1})\ p(2;2)\ p(2;3) \\ p(1;1)\ p(1;2)\ p(1;\hat{3}) \tag{3.24}$$

Two y-values, namely, 3 and 4 are mapped to $x = 2$, which gives $w(2) = p(4;\hat{2}) + p(3;\hat{2})$ while $w(1) = P(2;\hat{1})$ and $w(3) = p(1;\hat{3})$. Further

$$\hat{C} = \sum_{i=1}^{3} w(i). \tag{3.25}$$

We show next that the maximum estimation information or the maximum capacity is in general greater and never smaller than the maximum expected mutual information, the channel capacity.

Theorem 3.1 *Consider the model family $\{p(y;x) = p(y|x)\}$, where $\{p(y|x)\}$ is a set of conditionals such that $\sum_y p(y|x) = 1$. Let $w^*(x)$ be the prior which defines the channel capacity $I_{w^*}(Y;X)$. Then*

$$\log \hat{C} \geq I_{w^*}(Y;X). \tag{3.26}$$

Proof Let $p(y,x) = p(y|x)w^*(x)$ and let $p(y) = \sum_x p(y|x)w^*(x)$ denote the marginal. Then

$$p(y;\hat{x}(y)) \geq p(y|x), \tag{3.27}$$

for all x and y, with equality only when $p(y;x)$ is uniform. Further

$$\log \hat{C} = \log 1/\hat{p}(y) - \log 1/p(y;\hat{x}(y)) = \sum_{x,y} p(x,y) \log \hat{C} \tag{3.28}$$

$$= \sum_{x,y} p(x,y) \log 1/\hat{p}(y) - \sum_{x,y} p(x,y) \log 1/p(y;\hat{x}(y)) \tag{3.29}$$

$$= \sum_{y} p(y) \log 1/\hat{p}(y) + \sum_{x,y} p(x,y) \log p(y;\hat{x}(y)), \tag{3.30}$$

$$I_{w^*}(Y;X) = \sum_{y} p(y) \log 1/p(y) + \sum_{x,y} p(x,y) \log p(y|x), \tag{3.31}$$

$$\log \hat{C} - I_{w^*}(Y;X) = \sum_{y} p(y) \log \frac{p(y)}{\hat{p}(y)} + \sum_{x,y} p(x,y) \log \frac{p(y;\hat{x}(y))}{p(y|x)} \geq 0 \tag{3.32}$$

by the noise-free coding theorem and the *ML* estimator. The equality holds only when the distributions $p(y|x)$ for all x are identical and uniform.

We point out that except in the rare pathological case of equality the maximum estimation information cannot be reached by any prior, proper or improper, which appears to contradict the channel capacity. The explanation is that the representation $\hat{p}(y^n)$ of the family of distributions is fundamentally different from Shannon's mixture $\sum_{x^n} w(x^n) p(y^n|x^n)$ with any prior. In fact, it is better, giving a larger capacity and estimation information.

Example The channel capacity per symbol for the symmetric binary channel $P(1|0) = P(0|1) = q < 1/2$ was given in [4] as $I_{w^*}(Y;X) = 1 - h(q)$, where $w^*(0) = 1/2$ and $h(q)$ is the binary entropy. To calculate the maximum capacity for the same two distributions on $A = B = \{0,1\}$, namely $P(0|0) = P(1|1) = p = 1 - q$, we have $\hat{C}_1 = 2p$ and $\log \hat{C}_1 = 1 - \log 1/p$. Further, $\log 1/p = h(q) - q(\log 1/q - \log 1/p)$, and $\log 1/p < h(q)$ for $q < p$. This implies $1 - \log 1/p > 1 - h(q)$. Extended to the set of all binary strings of length n the capacities are $I_{w^*}(Y^n; X^n) = n(1 - h(q))$ and $\log \hat{C}_n = n(1 - \log 1/p)$, the first smaller than the second, as claimed by the theorem.

We conclude this subsection with a brief discussion of the channel coding and the role of the maximum capacity, which puts it in a different perspective. Let $\{\bar{x}(j) : j = 1, 2, \ldots, M\}$ be a subset of the set of all 2^n strings. Each string $\bar{x}(j)$ with the conditional per symbol channel probabilities p and q defines the binomial distribution $p(y^n|\bar{x}_j) = \binom{n}{m} p^{n-m} q^m$, where m denotes the Hamming distance $m = |y^n - \bar{x}(j)| = \sum_i |y_i - \bar{x}_i(j)|$. Let $\Lambda_m = \{B_j\}$ denote a partition of the set of all strings of length n with $\bar{x}(j)$ at the center of B_j with the probability $p(\bar{x}(j)|\bar{x}(j)) = p^n$. For finite strings consider the problem:

For each n and positive number ϵ find the maximum number $\hat{N}_{n,\epsilon}$ of centers and the partition they define such that the probability of all equivalence classes satisfies

$$P(B_j) = \sum_{y^n \in B_j} p(y^n|\bar{x}(j)) \geq 1 - \epsilon \tag{3.33}$$

for all $j \leq \hat{N}_{n,\epsilon}$, or, equivalently, $\min_j P(B_j) \geq 1 - \epsilon$.

For each n the maximum number $\hat{N}_{n,\epsilon}$, as a function of ϵ, grows and shrinks with ϵ, reaching its absolute maximum 2^n at $\epsilon = 1 - p^n$. It satisfies the inequalities

$$2^{n(1-h(q))} < \hat{N}_{n,\epsilon} \leq \hat{C}_{n,\epsilon}/(1-\epsilon), \qquad (3.34)$$

where $\hat{C}_{n,\epsilon} = \sum_j P(B_j)$. We call $\log \hat{C}_{n,\epsilon}$ the maximum capacity under the constraint that the worst case error probability is ϵ. It is between the channel capacity $n(1 - h(q))$ and the maximum capacity $n\log(2p) < n(1 - q\log e)$.

The probabilities $P(B_j)$ in the optimal partition, which minimize the worst case error probabilities, are as equal as they can be. It is clear that they must be so for otherwise we could take a string from the equivalence class where the error probability is smallest and put it into a class of the worst case error probability. This would increase $P(B_j)$ and decrease the worst case error probability, which is a contradiction.

In this framework the great result of Shannon's is that while the channel capacity gives a lower bound $2^{n(1-h(q))}$ of $\hat{N}_{n,\epsilon}$ for any ϵ, both $(1/n)\log \hat{C}_{n,\epsilon}$ and $(1/n)\log \hat{N}_{n,\epsilon}$ converge to $1 - h(q)$ when $\epsilon \to 0$ and $n \to \infty$. In addition he proved by ingenious random coding arguments the converse statement that there is a code with which any error probability can be reached with n large enough. This is clearly relevant for channel coding where the strings to be encoded can be long. The same holds even when we strengthen the restriction in the problem above by adding the requirement that the equivalence classes consist of strings such that their Hamming distance from the center grows from zero to a maximum while keeping the sum of their conditional probabilities above $1 - \epsilon$. The difficult problem is to find a practicable means to construct the collection of the centers.

Once we see that the channel capacity is a special case of the maximum capacity, the analogous general framework applies to estimation. The main difference is that in estimation the amount of data is often not large enough to justify asymptotic solutions, and instead of the channel capacity the important capacity is the maximum capacity together with its smaller restrictions under added conditions appropriate for estimation, as we shall show in the second part of the book.

Part II

Estimation

4 Modeling problems

The basic modeling problem begins with a set of observed data $y^n = \{y_t : t = 1, 2, \ldots, n\}$, generated by some physical machinery, where the elements y_t may be of any kind. Since no matter what they are they can be encoded as numbers we take them as such, i.e. natural numbers with or without the order if the data come from finite or countable sets, and real numbers otherwise. Often each number y_t is observed together with others $x_{1,t}, x_{2,t}, \ldots$, called *explanatory* data, written collectively as a $K \times n$ matrix $X = \{x_{i,j}\}$, and the data then are written as $y^n | X$. It is convenient to use the terminology "variables" for the source of these data. Hence, we say that the data $\{y_t\}$ come from the variable Y, and the explanatory data are generated by variables X_1, X_2, and so on.

In physics the explanatory data often determine the data y^n of interest, called a "law," but not so in statistical problems. Although by taking sufficiently many explanatory data we may also fit a function to the given set of observed data, but this is not a "law," since if the same machinery were to generate additional data $y_{n+1}, x_{1,n+1}, x_{2,n+1}, \ldots$ the function would not give y_{n+1}. This is the reason the objective is to learn the statistical properties of the data y^n, possibly in the context of the explanatory data. The statistical properties are represented by distributions, called *models*. We consider the data to consist of two parts. The first is what we may call intuitively "noise" and which we can do nothing about, but the second is the useful information bearing part, which we represent by the models and which is what we want to learn. When the data string is actually separated into the two parts, the noise and the useful information bearing part, the process is called "denoising," but in estimation we settle for less, just finding the statistical regular features represented by a model.

4.1 Models

We consider two general classes of models which we fit to data. The first class consists of parametric models of the type

$$\mathcal{M}_s = \{f(y^n|X(s);\theta,s) : \theta = \theta_1,\ldots,\theta_{k_s} \in \Omega^{k_s} \subset R^{k_s}\}, \tag{4.1}$$

which are either probability functions for quantized data or probability density functions (pdfs) when the data are considered to be real numbers. There are two kinds of parameters: a structure parameter s and real-valued parameters $\theta = \theta_1,\ldots,\theta_{k_s}$, whose number depends on the structure. The structure parameter typically selects the explanatory variables that we think most strongly affect the values of the variable Y of interest, and hence the *structure* is simply a subset of the models, defined by the real-valued parameters selected by the structure index. Often for each explanatory variable there is one real-valued parameter, like in linear regression, but not always. In curve fitting the data consist of the pairs (y_i, x_i) so that there is just one explanatory variable X, the argument of the fitted function, but if we fit polynomials there will be more than one parameter, namely, the parameters consisting of the coefficients of the powers of x_i.

For the most part in this book we consider structures defined by the first k explanatory variables $X_1, X_2, \ldots, X_k$ with the parameters $\theta = \theta_1,\ldots,\theta_k$, and $k \in N_K = \{1, 2, \ldots, K\}$ for $K \leq n$, and the number of parameters k is the only structure variable of interest. The derivations of the main results are the same whether we have general structures or structures that are determined by k. Moreover, with a fixed set of the explanatory data X we may drop the symbol X. If we keep n fixed to further simplify the notation we write $y^n = \mathbf{y}$, and we then consider the two model classes

$$\mathcal{M}_k = \{f(\mathbf{y};\theta) : \theta \in \Omega \subset R^k, k \leq K\}, \tag{4.2}$$

$$\mathcal{M} = \bigcup_{k \geq 1} \mathcal{M}_k, \tag{4.3}$$

where $f(\mathbf{y};\theta)$ is also written as $f(\mathbf{y};\theta,k)$, when the number of parameters needs to be indicated.

The model classes can be constructed in a number of ways, which raises the question of whether there is an optimal model class which would include an optimal model. Traditionally the whole issue is often avoided by the naive assumption that the data have been generated by a "true" distribution, which of course does not help at all. It is clear that when picking the class we must take into account the general type of the data and use all existing prior knowledge in the selection – not merely in the Bayesian sense of picking a prior distribution for the parameters. As we shall discuss in Appendix A the optimal selection is non-computable, even when the models are restricted to the computable distributions, and no adequate theory can exist for it. It is true that we can consider a collection of the above type of model classes as a kind of superclass, and apply our theory. However, that leaves us with the same problem, the optimal selection of the collection, which is non-computable, and so on. For these reasons we build our theory of estimation on a given model class somehow selected.

Our terminology "parametric models" is different from the traditional, and the full class of models $\mathcal{M}$ includes many of the so-called non-parametric models such as histograms, which actually can be described as a separate class of data, i.e., data that are given in the form of frequency counts rather than in terms of the numbers y_t. If more general non-parametric models can be fitted to data at all they must be defined by an algorithm – or by parameters as we do – and we do not discuss the estimation of general non-parametric models.

In traditional statistics only the estimation of the real-valued parameters is called "estimation" while the estimation of their number or structure is called "model selection." The reason for this is that the two tasks are done in entirely different manners: the estimation of the real-valued parameters is based on a rudimentary theory while no common theory exists for the "model selection." Rather, the task is done by numerous criteria often suggested by the imagined properties of the "truth."

Since finite sets and open sets with their closures are the only measurable sets we need we do not use the terminology "measurable sets" and "measurable functions" appropriate in probability theory. In fact, in order to simplify the presentation we often use the arguments for probability functions with the understanding that the same arguments apply to pdfs which can be quantized to give probabilities on quantized data.

Corresponding to the two model classes we have sets $\mathcal{F}_k = \{\bar{\theta}(\cdot)\}$ and $\mathcal{F} = \{\bar{\theta}(\cdot), \bar{k}(\cdot)\}$ of *estimator* functions

$$\bar{\theta}(\cdot) : \mathbf{y} \mapsto \bar{\theta}(\mathbf{y}) \in R^k, \tag{4.4}$$

$$\bar{\theta}(\cdot), \bar{k}(\cdot) : \mathbf{y} \mapsto \bar{\theta}(\mathbf{y}), \bar{k}(\mathbf{y}). \tag{4.5}$$

Although the estimators are functions of both $\mathbf{y}$ and X the important thing is the dependence of the estimated parameters on $\mathbf{y}$ only, just as the case was with the models themselves, and we drop the symbol X in the notation of the estimated parameter values: $\bar{\theta}(\mathbf{y}) = \bar{\theta}(\mathbf{y}, X)$ and $\bar{k}(\mathbf{y}) = \bar{k}(\mathbf{y}, X)$. The estimator functions are occasionally also denoted $\bar{\theta}$ and $\bar{\theta}, \bar{k}$, but these are not to be confused with parameter values.

There is a complication in the model classes discussed, namely, that the parameter values θ, k determine a model $f(\mathbf{y}; \theta, k)$, where we now indicate the number of parameters, but the converse is not true, because only the non-zero parameters count, and both $\theta_1, \ldots, \theta_k$ and $\theta_1, \ldots, \theta_k, 0, \ldots, 0$ determine the same model. As a consequence the so-called "likelihood" $f(\mathbf{y}; \theta, k)$ is maximized by the largest number of non-zero parameters, which creates a serious problem. After all, the maximizing parameter values in the position of the 0s cannot hurt and they may increase the maximum, which has led to the misunderstanding that the more parameters there are in the model the better it is because it is closer to the "truth," and the "truth" obviously is not simple. Also, the real-valued parameters and their number are linked in that any selection of the real-valued non-zero parameters defines their number and the model, while the number k alone determines only a class of models, rather than a specific model.

To conclude this section we give two examples, which illustrate the source of large numbers of parameters.

1. Linear regression: Span the space of "smooth" signals $\hat{\mathbf{y}} = \hat{y}^n = \hat{y}_1, \hat{y}_2, \ldots, \hat{y}_n$ by basis vectors $\{\mathbf{x}(i)\}$, the rows of the matrix X, and write the observed signal $\mathbf{y}$ as

$$\mathbf{y} = \hat{\mathbf{y}} + \mathbf{e}, \tag{4.6}$$

$$\hat{\mathbf{y}} = \sum_{i=1}^{k} \theta_i \mathbf{x}(i). \tag{4.7}$$

Take the family of models $f(y_t|\hat{y}_t, \sigma^2)$ as normally distributed density functions of mean $\hat{y}_t$ and variance σ^2, or, equivalently, for e_t a similarly distributed density function of mean 0 and variance σ^2. If we just need to consider a class of normal distributions without the need to specify the parameters we often call them simply "gaussians." These are extended to sequences $f(\mathbf{y}|\hat{\mathbf{y}}, \sigma^2, k)$ by the product.

2. Markov models: Take the explanatory variables as states $s_t = F(y^t)$, and with suitably defined conditionals $f(y_t|s_{t-1}, \theta)$ we get a *Markovian* model for $\mathbf{y}$ by the product of the conditionals.

We discuss both model classes later.

4.2 General comments on estimation

The basic difficulty with estimation is the fact that each data sequence can be a sample from all the distributions in the family $\mathcal{M}$, which creates the problem of how to prefer one over another. The traditional way to approach this problem is to imagine that perhaps instead of the "truth" a model $f(\mathbf{x}; \theta, k)$, defined by a parameter θ, has generated the data, and the objective is to estimate it. Instead, our objective is to estimate statistical properties of the data, where the properties are represented by the models in a selected class, say $\mathcal{M}_k$ for k fixed, or, more generally, $\mathcal{M}$ in which k, too, is to be found.

For each sequence $\mathbf{x}$ the models defined by the parameters (θ, k) are not its properties. Instead its properties are the models $f(\mathbf{y}; \bar{\theta}(\mathbf{x}), \bar{k}(\mathbf{x}))$ evaluated at the observed data $\mathbf{x}$ and defined by the estimates $(\bar{\theta}(\mathbf{x}), \bar{k}(\mathbf{x}))$, obtained by the various estimators $(\bar{\theta}(\cdot), \bar{k}(\cdot))$. Hence, the process is analogous to measuring a physical property such as the weight of a material object. The objective then is to study the properties of the estimators, like the instruments by which physical properties are measured. It is possible to consider different representations of the statistical properties of data such as functions of the members in these classes, a familiar example being the Bayesian mixture $\int f_w(\mathbf{x}; \theta, k) w(\theta) d\theta$, where $w(\theta)$ is a prior. We consider such representations only briefly, because they do not involve estimation of specific parameters.

We illustrate the fundamental ideas behind estimation in a simple example. Consider the set of all binary strings of length n, and the estimator for the mean, the sum of 1s in the string divided by n. There are $n+1$ possible values of the sum, each defining an equivalence class in the partition of the 2^n strings under the estimator function. This is the constraint on the strings that we are interested in, and having observed a string x^n we wish to learn the constraint. For a particular sum value, say m, there are $\binom{n}{m}$ strings in the pile where x^n belongs. The pile with the most strings is for $m = n/2$, and the size shrinks symmetrically as m gets further away from $n/2$. The piles behave somewhat like piles of sand under gravity, although there is no gravity here. It is an amazing fact that m divided by n approaches the normally distributed probability curve, more and more closely as n grows, which is an instance of the CLT. Somehow the normal curve allows for the most efficient packing of the strings. As a matter of fact, if we try to alter the pile sizes from $\binom{n}{m}$ by some other estimator we would find that there would be strings left over, which could not be placed in any of the piles. This is why and how we can learn the constraints in any string. A particular string x^n neither has a "true" constraint nor is it hiding in some pile for us to find. It is simply a matter of the number of the strings and the estimators which impose constraints on the strings. The best strategy to learn the constraint is to find the pile with the most strings where x^n lies, which can be found by the arithmetic mean, and which amounts to the ML estimation. Also, each pile is "locally" generated, rather than forced onto us by the other piles. This too turns out to be important as we shall see.

There are of course much more intricate constraints than just the mean, especially those involving the structure such as the number of parameters, and the "piles" we need to consider must be created by these constraints for us to be able to learn them. For instance, the ML estimate of the number of parameters is n, and all the strings belong to the single pile. Clearly, this teaches us nothing, because there is no constraint reflecting the number of non-zero parameters. This again kills the sometimes offered view that the best strategy is to fit the model with the maximum number of parameters.

The reader may wonder why we consider the classes of models simply as a means to represent statistical properties rather than a set where the "true" data generating distribution is hiding, which is the traditional view inherited from physics. The reason for our view appears to be that the usual "true laws" in physics differ from those in statistics in an important respect. If we try to verify Ohm's law in a closed electric circuit it does not matter if we measure the current and voltage ratio 10 times or 100 times or any number of times, we cannot detect any deviation from the "law": the current is proportional to the voltage by a fixed constant of proportionality, and the law is "true." The same is not the case in statistics. Whatever statistical property we want to verify, such as a distribution from its samples, the result depends strongly on the number of measurements, i.e. the data points we observe. Hence, if we are hunting for the "truth" we must admit that there are a number of models that are just as good in any reasonable sense as the "true" distribution we are after, and there is no way to select a preferred one. One may then argue that if we had an infinite number of data our method would find the true distribution, but such a theory would not be a theory of statistics; it would belong to probability theory and mathematics. Finally, a strong case can be made that no model class captures the "truth," which makes the whole notion useless and in some cases even harmful. This tends to happen if we insist in investigating something that does not exist and we lose interest in things and events that do exist. This happens with analyses of the almost sure behavior of estimators on infinite strings, which unless used with caution can lead to grossly distorted conclusions. We discuss this further in Section 5.2. Other examples are in hypothesis testing, where the assumption that the hypotheses are "true" has misguided the entire field by generating problems which do not exist and distorting rational solutions to problems that do exist.

4.3 Maximum capacity

The foundation of this theory is based on the premise that a good estimator should separate the different estimated models well, regardless of whether we imagine the data as having been generated by some model or not. An estimator's separation property can be measured by the amount of the probability mass that a model, selected by an estimator, is capable of assigning to the observed data. We know from coding theory that an equivalent measure is the negative logarithm of the probability, which gives the number of bits with which the data can be described and stored. This has a strong intuitive appeal, because nonsensical data require a long code length and are difficult to remember and understand. Also, if the probability is large the data have little surprise and the prediction of the data points has a small probability of errors. We regard this criterion as a postulate in the theory of estimation we develop in the belief that regardless of the intended use of the model for best performance the estimator should maximize the probability on the data. However, this poses a problem, because no distribution exists that gives the largest probability to all the data, and we seem to have reached an impasse. This has presumably stymied all previous attempts to apply such a criterion.

The way out is to put restrictions on the estimators, the first of which is that the estimators must be determined by the model class to prevent tailoring to the data. This still leaves a lot of such "fair" estimators from which to select the best. We do it by requiring that an estimator must satisfy any of the following three equivalent requirements:

1. it must maximize the capacity or estimation information;
2. it must satisfy necessary conditions for maximizing the probability on the data;
3. it must be in accord with the complete MDL principle.

We discuss the first property for models in both of the classes $\mathcal{M}_k$ and $\mathcal{M}$ in this section and the other two cases in the subsequent sections. We start with the estimation of the real-valued parameters of a fixed known number, and then discuss the general case where also the number of parameters is to be estimated.

Consider an estimator $\bar{\theta}(\cdot)$, where $\bar{\theta}(\mathbf{y})$ has k components. It defines the density function and model $f(y; \bar{\theta}(\mathbf{x}), k)$ from an observed sequence $\mathbf{x}$, which may be extended to sequences $f(\mathbf{y}; \bar{\theta}(\mathbf{x}), k)$. Although this looks like a good candidate to give us the probability to maximize, namely

$$\max_{\theta} f(\mathbf{x}; \theta, k), \qquad (4.8)$$

which defines the ML estimator $\hat{\theta}(\cdot)$, it falls short of the postulate we want. First, it does not define a density function, for $f(\mathbf{x}; \hat{\theta}(\mathbf{x}), k)$ does not integrate to unity, and in addition and most importantly it cannot be applied to the estimation of the number of parameters or structures, which would have to be done by ad hoc means with a meager theory of estimation as the result.

As we did for the maximum capacity in Part I we can construct another distribution for each estimator function:

$$\bar{f}(\mathbf{x}; k) = f(\mathbf{x}; \bar{\theta}(\mathbf{x}), k)/\bar{C}_k, \qquad (4.9)$$

$$\bar{C}_k = \int f(\mathbf{y}; \bar{\theta}(\mathbf{y}), k) d\mathbf{y} \qquad (4.10)$$

$$= \int_B g(\bar{\theta}; \bar{\theta}, k) d\bar{\theta} < \infty, \qquad (4.11)$$

$$g(\bar{\theta}; \theta, k) = \int_{\mathbf{y}:\bar{\theta}(\mathbf{y})=\bar{\theta}} f(\mathbf{y}; \theta, k) d\mathbf{y}, \qquad (4.12)$$

where the range of the integration B is selected so as to make the integral finite. We discuss in Chapter 8 an example of its optimal selection.

Similarly for the case when the number of the parameters is also to be estimated the estimator $\bar{\theta}(\cdot), \bar{k}(\cdot)$ defines the density function

$$\bar{f}(\mathbf{x}) = \bar{f}(\mathbf{x}; \bar{k}(\mathbf{x}))/\bar{C}, \qquad (4.13)$$

$$\bar{C} = \int \bar{f}(\mathbf{y}; \bar{k}(\mathbf{y})) d\mathbf{y}. \qquad (4.14)$$

We now redefine the *maximum capacity*, introduced in Part I, for the class $\mathcal{M}_k$

$$\log \hat{C}_k = \max_{\bar{\theta}(\cdot)} \log \bar{C}_k. \qquad (4.15)$$

We call the maximizing estimator $\hat{\theta}(\cdot)$ the *maximum capacity* estimator, which agrees with the ML estimator. The reason for the different name is that it generalizes to the estimators for both the real-valued parameters and their number, which the ML estimator does not. We also see that it is defined without any data in accordance with the requirement of the postulate that an estimator be determined by the model class. It also equals the maximum amount of information an estimator can obtain about the random variable $\mathbf{X}$ with the distribution $\hat{f}(\mathbf{x}; k)$, which we called the *maximum estimation information* in Part I, i.e. the logarithm of the ratio of $f(\mathbf{x}; \hat{\theta}(\mathbf{x}), k)$ and $\hat{f}(\mathbf{x}; k)$

$$\log \frac{f(\mathbf{x}; \hat{\theta}(\mathbf{x}), k)}{\hat{f}(\mathbf{x}; k)} = \log \hat{C}_k \geq 0, \qquad (4.16)$$

Modeling problems

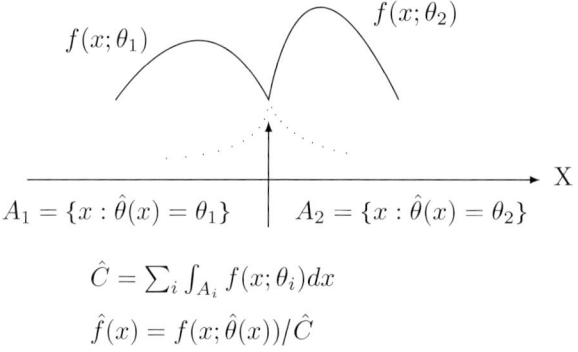

Figure 4.1 Maximum capacity for two models.

with the last inequality because for all θ

$$\int f(\mathbf{y}; \hat{\theta}(\mathbf{y}), k) d\mathbf{y} \geq \int f(\mathbf{y}; \theta, k) d\mathbf{y} = 1. \tag{4.17}$$

As any function the estimator $\hat{\theta}(\cdot)$ defines an equivalence relation for the data sequences: $\mathbf{y} \equiv \mathbf{x}$ if $\hat{\theta}(\mathbf{y}) = \hat{\theta}(\mathbf{x})$. Let

$$g(\hat{\theta}; \theta, k) = \int_{\mathbf{y} \equiv \mathbf{x}} f(\mathbf{y}; \theta, k) d\mathbf{y} \tag{4.18}$$

be the density function on the ML estimates, induced by the model $f(\mathbf{y}; \theta, k)$. It further induces the conditional density function for the data sequences $\mathbf{y} \equiv \mathbf{x}$ by

$$f(\mathbf{y}|\hat{\theta}(\mathbf{x}), k) = f(\mathbf{y}; \hat{\theta}(\mathbf{x}), k) / g(\hat{\theta}(\mathbf{x}); \hat{\theta}(\mathbf{x}), k). \tag{4.19}$$

We can then also write

$$\hat{f}(\mathbf{x}; k) = f(\mathbf{x}|\hat{\theta}(\mathbf{x}), k)\hat{g}(\hat{\theta}(\mathbf{x}); k), \tag{4.20}$$

$$\hat{g}(\hat{\theta}; k) = g(\hat{\theta}; \hat{\theta}, k) / \hat{C}_k, \tag{4.21}$$

$$\hat{C}_k = \int g(\hat{\theta}; \hat{\theta}, k) d\hat{\theta}. \tag{4.22}$$

In summary, the capacity is defined by the estimator function $\hat{\theta}(\cdot)$, which assigns the maximum amount of probability mass or density to the equivalence classes: $\mathbf{x} \equiv \mathbf{y}$ if $\hat{\theta}(\mathbf{x}) = \hat{\theta}(\mathbf{y})$. The density or probability function $\hat{g}(\hat{\theta}; k)$ on these equivalence classes is particularly important.

Figure 4.1 shows an example of the maximum capacity for a class of two models.

Example We calculate the maximum capacity and its asymptotic approximation for the Bernoulli class $\mathcal{B}$ with $P(x=0) = \theta$ as the parameter. We have

$$P(x^n; \theta) = \theta^{n_0}(1-\theta)^{n-n_0}, \tag{4.23}$$

where n_0 denotes the number of 0s in x^n. The distribution $\hat{P}(x^n) = \hat{f}(\mathbf{x}; k)$, defined by the maximum capacity, is given by

$$\hat{P}(x^n) = \frac{P(x^n; \hat{\theta})}{\sum_m \binom{n}{m}\left(\frac{m}{n}\right)^m \left(\frac{n-m}{n}\right)^{n-m}}, \tag{4.24}$$

where $\hat{\theta} = \hat{\theta}(x^n) = n_0/n$ and $P(x^n; \hat{\theta}) = \hat{\theta}^{n_0}(1-\hat{\theta})^{n-n_0}$. The distribution of the ML estimates is the binomial

$$P(\hat{\theta}; \theta) = \binom{n}{n_0} P(x^n; \theta). \tag{4.25}$$

The distribution $p(y^n|\hat{\theta}) = \binom{n}{n_0}^{-1}$ is seen to be very different from $P(y^n; \hat{\theta})$. The maximum capacity is given by

$$\log \hat{C} = \log \sum_m \binom{n}{m}\left(\frac{m}{n}\right)^m \left(\frac{n-m}{n}\right)^{n-m}. \tag{4.26}$$

To evaluate the capacity we use the important Stirling's approximation formula in the form refined by Robbins:

$$\ln n! = (n+1/2)\ln n - n + \ln\sqrt{2\pi} + R(n), \tag{4.27}$$

where $1/12(n+1) \leq R(n) \leq 1/12n$.

Then

$$\ln \binom{n}{m} \approx n h(m/n) - \frac{1}{2}\ln[(m/n)(n-m)/n] - \frac{1}{2}\ln n - \ln\sqrt{2\pi}, \tag{4.28}$$

where $h(p)$ is the binary entropy at p. This gives

$$\binom{n}{m}\left(\frac{m}{n}\right)^m \left(\frac{n-m}{n}\right)^{n-m} \cong \frac{1}{\sqrt{2\pi n}}[(m/n)(n-m)/n]^{-1/2}. \tag{4.29}$$

The right hand side in the sum

$$\sum_m \binom{n}{m}\left(\frac{m}{n}\right)^m \left(\frac{n-m}{n}\right)^{n-m} \cong \sqrt{n/(2\pi)} \sum_m [(m/n)(n-m)/n]^{-1/2}\frac{1}{n} \tag{4.30}$$

with step length $1/n$ can be approximated by the Riemann integral

$$\sqrt{n/(2\pi)} \int_0^1 \frac{1}{\sqrt{p(1-p)}}dp = \sqrt{n\pi/2}. \tag{4.31}$$

The integral of the square root of the Fisher information $J(p) = 1/(p(1-p))$ is Dirichlet's integral with the value π. Finally, the maximum capacity is given by

$$\log \hat{C} = \frac{1}{2} \log \frac{n\pi}{2} + o(1). \tag{4.32}$$

The added term $o(1)$, which goes to zero as $n \to \infty$, takes care of the errors made by the application of Stirling's formula and the errors in the approximation of the sum by the integral.

For the class $\mathcal{M}$, when the number of parameters is also estimated, the capacity is defined quite similarly:

$$\log \hat{C} = \max_{\bar{\theta}(\cdot), \bar{k}(\cdot)} \log \sum_k \int_{\bar{k}(\mathbf{y})=k} \bar{f}(\mathbf{y}; k) d\mathbf{y} \tag{4.33}$$

$$= \log \sum_k \int_{\hat{k}(\mathbf{y})=k} \hat{f}(\mathbf{y}; k) d\mathbf{y}, \tag{4.34}$$

where $\hat{k}(\mathbf{y})$ is the value of k that maximizes $\hat{f}(\mathbf{y}; k) = f(\mathbf{y}; \hat{\theta}(\mathbf{y}), k)/\hat{C}_k$ rather than just the numerator or the likelihood $f(\mathbf{y}; \hat{\theta}(\mathbf{y}), k)$. In fact, the ML estimator gives the maximum number K of parameters and assigns the probability mass unity to the single equivalence class, which is smaller than $\hat{C}$.

We also call the estimator $\hat{\theta}(\mathbf{y}), \hat{k}(\mathbf{y})$ the *maximum capacity* estimator. The maximum capacity estimator assigns the maximum amount of probability mass,

$$P(\hat{k}; \hat{k}) = \int_{\hat{k}(\mathbf{y})=\hat{k}} \hat{f}(\mathbf{y}; \hat{k}) d\mathbf{y}, \tag{4.35}$$

to equivalence classes defined by $\mathbf{x} \equiv \mathbf{y}$ if $\hat{k}(\mathbf{x}) = \hat{k}(\mathbf{y})$, which permits the maximum separation. The maximum capacity estimator also maximizes the estimation information

$$\log \hat{C} = \log \frac{\hat{f}(\mathbf{x}; \hat{k}(\mathbf{x}))}{\hat{f}(\mathbf{x})}, \tag{4.36}$$

$$\hat{f}(\mathbf{x}) = \frac{\max_k f(\mathbf{x}; \hat{\theta}(\mathbf{x}), k)/\hat{C}_k}{\hat{C}}. \tag{4.37}$$

Further

$$\hat{f}(\mathbf{x}) = \frac{\hat{f}(\mathbf{x}; \hat{k}(\mathbf{x}))}{\hat{C}} = \hat{f}(\mathbf{x}|\hat{k}(\mathbf{x}))\hat{P}(\hat{k}(\mathbf{x})), \quad (4.38)$$

$$\hat{f}(\mathbf{y}|\hat{k}(\mathbf{x})) = \frac{\hat{f}(\mathbf{y}; \hat{k}(\mathbf{x}))}{\hat{P}(\hat{k}(\mathbf{x}), \hat{k}(\mathbf{x}))}, \quad (4.39)$$

$$\hat{P}(\hat{k}) = \frac{P(\hat{k}; \hat{k})}{\hat{C}}, \quad (4.40)$$

$$\hat{C} = \sum_{k} P(k; k). \quad (4.41)$$

Indeed

$$\int \hat{f}(\mathbf{y}; \hat{k}(\mathbf{y})) d\mathbf{y} = \sum_{k} \int_{\mathbf{y}:\hat{k}(\mathbf{y})=k} \hat{f}(\mathbf{y}; k) d\mathbf{y} = \sum_{k} P(k; k) = \hat{C} \quad (4.42)$$

so that $\hat{f}(\mathbf{y})$ is a distribution.

4.4 Necessary conditions for optimality

We now study another way to derive the same optimal estimators $\hat{\theta}(\mathbf{x}), \hat{k}(\mathbf{x})$. It illustrates different properties of these estimators, which will be useful later to obtain further assurance of the probability maximizing criterion we have selected as the postulate. Instead of invoking the somewhat abstract concepts like maximum capacity and maximum estimation information to define optimality as in the preceding section, we appeal to the postulate itself.

Theorem 4.1 *Let the capacity* $\log \bar{C}$ *be finite defining the probability of all points* $\mathbf{x}$ *or the density at them*

$$\bar{f}(\mathbf{x}) = \frac{f(\mathbf{x}; \bar{\theta}(\mathbf{x}), \bar{k}(\mathbf{x}))/\bar{C}_{\bar{k}(\mathbf{x})}}{\bar{C}}. \quad (4.43)$$

A necessary condition for estimators $(\bar{\theta}(\cdot), \bar{k}(\cdot))$ *to maximize the ratio at* $\mathbf{x}$ *is to maximize the numerator. The only estimator satisfying the condition for all* $\mathbf{x}$ *is* $(\hat{\theta}(\mathbf{x}), \hat{k}(\mathbf{x}))$.

Proof The proof is based on repeated use of the following immediate inequality, which we state as a lemma for later use.

Lemma 4.1 *For positive numbers a, b, and c, $a < c$,*

$$\frac{a}{c} < \frac{a+b}{c+b}, \tag{4.44}$$

and if $\hat{b}$ maximizes the numerator the ratio is also maximized.

To see how this determines the necessary conditions for the optimality of the ML estimates for a given number of (non-zero) parameters and discrete data put $a = f(\mathbf{x}; \bar{\theta}(\mathbf{x}), k)$, $c = \bar{C}_k < \infty$, and $b = f(\mathbf{x}; \hat{\theta}(\mathbf{x}), k) - f(\mathbf{x}; \bar{\theta}(\mathbf{x}), k)$. For $\hat{\theta}(\mathbf{x})$ the new ratio $f(\mathbf{x}; \hat{\theta}(\mathbf{x}), k)/(\bar{C}_k + b)$ is larger than $\bar{f}(\mathbf{x}; k)$, which means that $\bar{\theta}(\mathbf{x})$ is not optimal at $\mathbf{x}$ unless it is $\hat{\theta}(\mathbf{x})$.

For real-valued data this is obvious, because with one point change of $f(\mathbf{x}; \bar{\theta}(\mathbf{x}), k)$ the integrals $\bar{C}_k = \bar{C}_k + b$ are equal. Clearly, the inequality remains intact if we add $f(\mathbf{y} : \hat{\theta}(\mathbf{y}), k) - f(\mathbf{y}; \bar{\theta}(\mathbf{y}), k)$ at each $\mathbf{y}$ in a neighborhood of $\mathbf{x}$, and take b as the integral of the difference over this neighborhood.

We prove the theorem for the general estimators for discrete data. Suppose the condition is negated. It can happen in two ways:

1. $(\bar{\theta}(\mathbf{x}), \bar{k}) \neq (\hat{\theta}(\mathbf{x}), \bar{k})$;
2. $(\bar{\theta}(\mathbf{x}), \bar{k}) = (\hat{\theta}(\mathbf{x}), \bar{k})$ and $\bar{k}(\mathbf{x}) \neq \hat{k}(\mathbf{x})$.

In the first case, pick in Lemma 4.1 for discrete data $a = f(\mathbf{x}; \bar{\theta}(\mathbf{x}), \bar{k})$, $c = \bar{C}_k$, and $b = f(\mathbf{x}; \hat{\theta}(\mathbf{x}), \bar{k}) - f(\mathbf{x}; \bar{\theta}(\mathbf{x}), \bar{k})$. This gives the inequality

$$\bar{f}(\mathbf{x}; \bar{k}) < \frac{f(\mathbf{x}; \hat{\theta}(\mathbf{x}), \bar{k})}{\bar{C}_k + b}, \tag{4.45}$$

which is enough to show that $\bar{C} < \hat{C}$ and the non-optimality of $\bar{f}(\mathbf{x})$.

We need to show that the right hand side of (5.49) can be increased by replacing $\bar{k}$ by $\hat{k}$, which shows that neither side is optimal. Let $\tilde{C}_{\bar{k}} = \bar{C}_{\bar{k}} + b_1$. Then put in Lemma 4.1 $b = b_2 = f(\mathbf{x}; \hat{\theta}(\mathbf{x}), \hat{k})/\tilde{C}_{\bar{k}} - f(\mathbf{x}; \hat{\theta}(\mathbf{x}), \bar{k})/\bar{C}_{\bar{k}}$, $a = f(\mathbf{x}; \hat{\theta}(\mathbf{x}), \bar{k})/\bar{C}_{\bar{k}}$ and $c = \tilde{C}_{\bar{k}}$, which give

$$\bar{f}(\mathbf{x}; \bar{k}) < \frac{f(\mathbf{x}; \hat{\theta}(\mathbf{x}), \bar{k})}{\tilde{C}_{\bar{k}}} < \frac{\hat{f}(\mathbf{x}; \hat{\theta}(\mathbf{x}), \hat{k})}{\tilde{C}_{\bar{k}} + b_2}. \tag{4.46}$$

Since the left-most term is non-optimal neither is $\bar{f}(\mathbf{x})$.

The proof for generalization to real-valued data is similar to the first case.

4.4 Necessary conditions for optimality

A consequence of Theorem 4.1 is the following. The calculation of $\hat{C}$ can be difficult, especially when it involves the optimized structures rather than just the number of parameters. Provided we verify that $\hat{C}$ is finite and there are no common factors in the ratio, Theorem 4.1 makes the calculation unnecessary since it is enough to maximize the numerator $\hat{f}(\mathbf{x}; k)$ to get the optimal structure. Notice, however, that this is not enough for data compression, for which the entire function $\hat{f}(\mathbf{x})$ is needed. This is an important difference between data compression and estimation.

We discuss briefly the selection of the range of the integration B in the previous section. Write the range of integration as $B = B(\hat{\theta}(\mathbf{x}); \beta)$, where β is a hyperparameter, and the maximum capacity as $\log \hat{C}_{k,\mathbf{x}}$. By the postulate the hyperparameter should be optimized thus:

$$\hat{C}_{k,\mathbf{x}} = \min_{\beta} \int_{B(\hat{\theta}(\mathbf{x});\beta)} g(\hat{\theta}; \hat{\theta}, k) d\hat{\theta}. \qquad (4.47)$$

Indeed, although by Theorem 4.1 $\hat{C}_k$ should be maximized with respect to the estimator $\bar{\theta}(\cdot)$ for all ranges B, it should be minimized with respect to the range, because then $\hat{f}(\mathbf{x}; k)$ is maximized.

The justification of the ML estimator provided by the necessary condition is as follows. If the numerator of the ratio $\bar{f}(\mathbf{x}; k)$ at $\mathbf{x}$ is not maximized neither is the ratio itself, which is the probability criterion we have selected as the universal criterion. It then follows that the ML estimator $\hat{\theta}(\cdot)$ and the estimator $(\hat{\theta}(\cdot), \hat{k}(\cdot))$ are the only estimators which satisfy the necessary conditions for all $\mathbf{x}$. This clearly is the strongest achievable justification as far as the probability criterion is concerned. Traditionally the ML estimator has had mostly intuitive justification. The formal justification is restricted to the Cramér–Rao inequality, which for finite data holds only under strong conditions for the models and without them only asymptotically. We strengthen the Cramér–Rao inequality in Chapter 5.

Consider an estimator "fair," if it does not "favor" any particular set of data at the cost of reducing the performance on other data. All the traditional estimators for the real-valued parameters, such as least squares, the method of the moments, and the traditional ML estimators are "fair." The same is true of the estimators of the number of parameters resulting from optimization of the multitude of the existing criteria. If we take the necessary conditions for all data as a requirement for the optimality of estimators, then because of the uniqueness the estimator $\hat{\theta}(\cdot), \hat{k}(\cdot)$ could be said to satisfy sufficient conditions as well. We then call the maximum capacity estimator $\hat{\theta}(\cdot), \hat{k}(\cdot)$ and the model it defines *perfectly fair*. The other fair criteria mentioned except the ML estimator are not perfectly fair.

To summarize the two different but equivalent ways to derive the optimal estimators we rewrite the distributions they define as follows:

$$\hat{C}_k = \int_B \mathrm{d}\theta \int_{\hat{\theta}(\mathbf{y})=\theta} f(\mathbf{y};\hat{\theta}(\mathbf{y}),k)\mathrm{d}\mathbf{y} = \int_B \hat{g}(\hat{\theta},k)\mathrm{d}\hat{\theta}, \qquad (4.48)$$

$$\hat{f}(\mathbf{y};k) = \frac{f(\mathbf{y};\hat{\theta}(\mathbf{y}),k)}{\hat{C}_k}, \qquad (4.49)$$

$$\hat{f}(\mathbf{y}) = \frac{\max_k f(\mathbf{y};\hat{\theta}(\mathbf{y}),k)/\hat{C}_k}{\hat{C}} \Rightarrow \hat{k}(\mathbf{y}), \qquad (4.50)$$

$$\hat{C} = \sum_k \int_{\hat{k}(\mathbf{y})=k} f(\mathbf{y};\hat{\theta}(\mathbf{y}),k)\mathrm{d}\mathbf{y}/\hat{C}_k, \qquad (4.51)$$

where the structure (number of parameters) that maximizes $\hat{f}(\mathbf{y};k)$ rather than the likelihood $f(\mathbf{y};\hat{\theta}(\mathbf{y}),k)$ is written as $\hat{k}(\mathbf{y})$.

The model $\hat{f}(\mathbf{y};k)$ was introduced by Shtarkov [49] as a universal model for data compression; we discuss it further in the next section. The second model $\hat{f}(\mathbf{y})$ was introduced in my 2009 Shannon Lecture. We do not consider these as universal models applicable in general to problems outside of data compression. Rather, we regard $f(\mathbf{y};\hat{\theta}(\mathbf{x}),k)$ and $f(\mathbf{y};\hat{\theta}(\mathbf{x}),\hat{k}(\mathbf{x}))$ for the two model classes, respectively, as the models applicable to all data generated by a physical machinery, whose statistical behavior we hope is captured by these models. We call these "batch" models and discuss in the last chapter different random process models, which are appropriate for prediction and control.

We can now compare two models f_1 and f_2, defined by the estimators $\bar{\theta}_1(\cdot)$ and $\bar{\theta}_2(\cdot)$, respectively, by the size of the sets $\bar{A}_i = \{\mathbf{x} : \bar{\theta}_i(\cdot) \neq \hat{\theta}_i(\cdot)\}$ of guaranteed non-optimality. We say that f_1 is better than f_2 if $|\bar{A}_1| < |\bar{A}_2|$, because $\bar{A}_i$ is the set of strings $\mathbf{x}$ where the probability $f_i(\mathbf{x})$ can be increased. Clearly, the smaller this set is the better the model, and the more confident we are that we have learned well the statistical properties in the given string $\mathbf{x}$. For the ML estimator this set is empty.

This view of estimation, based on the postulate criterion, differs from the customary one in a fundamental way. Most importantly, any role of a "true" data generating model is gone. Even if we knew that a data set $\mathbf{x}$ is a sample of the distribution $f(\mathbf{x};\theta)$ for some k-dimensional parameter θ, we would have to dismiss it, because the estimated model $f(\mathbf{y};\hat{\theta}(\mathbf{x}))$ defined by $\hat{\theta}(\mathbf{x})$ describes the statistical property more sharply and better.

4.5 General and complete MDL principles

The principle of maximizing the probability as the criterion was actually introduced a long time ago. The problem that no coding is possible with the shortest code length for all data was avoided by deliberately leaving the definition of the entire model class vague. For a model class, where a portion consists of a set such as $\mathcal{M}_k$, the original MDL principle, which we now call the *incomplete* or *general* MDL principle, can be stated thus [37] (for an earlier more primitive version see [56]):

> Find a model with which the observed data and the model can be encoded with the shortest code length

$$\min_{\theta,k} \left[\log \frac{1}{f(\mathbf{x};\theta,k)} + L(\theta,k) \right], \qquad (4.52)$$

where $L(\theta,k)$ denotes the code length for the parameters defining the model. The only requirement for the code length of the optimizing parameters $L(\hat{\theta}(\mathbf{x}),\hat{k}(\mathbf{x}))$ is that they be decodable. As described below this does not require any prior.

The *complete* MDL principle differs from the general principle in the requirement that the code length is the negative logarithm of a probability or density, defined by the distribution $\hat{f}(\mathbf{x};k)$, (4.49), and in the general case where even the number of parameters is estimated by $\hat{f}(\mathbf{x})$, (4.50). Since both are determined by the model class its description or code length is common for all data and can be ignored. Both codes $\hat{f}(\mathbf{x};k) = f(\mathbf{x};\hat{\theta}(\mathbf{x}),k)/\hat{C}_k$ and $\hat{f}(\mathbf{x})$ are complete.

The necessary conditions imply that we can state the crucial property of the principle as follows:

No string's codeword can be shortened from

$$\log 1/\hat{f}(\mathbf{x};\mathcal{M}) = \log 1/f(\mathbf{x};\hat{\theta}(\mathbf{x}),\hat{k}(\mathbf{x})) + \log \hat{C}_{\hat{k}(\mathbf{x})} + \log \hat{C} \qquad (4.53)$$

by any choice of estimated parameters.

The complete code $\hat{f}(\mathbf{x};k)$ for a fixed k, defined by the capacity $\log \hat{C}_k$, was originally suggested by Shtarkov for data compression [49], and later it was extended to estimation as a universal model under the name normalized maximum likelihood (NML) model [2]. He defined it by asking for a universal code with code length closest to the ideal $\log 1/f(\mathbf{x};\hat{\theta}(\mathbf{x}),k)$ as the solution $\hat{f}(\mathbf{x};k) = f(\mathbf{x};\hat{\theta}(\mathbf{x}),k)/\hat{C}_k$ to the minmax problem

$$\min_q \max_\mathbf{x} \log \frac{f(\mathbf{x};\hat{\theta}(\mathbf{x}),k)}{q(\mathbf{x})}. \qquad (4.54)$$

The minmax procedure actually fails to provide a formalized justification for the ML estimator or the code it defines, because the minmax argument applies even to codes defined by estimators $\bar{\theta}(\cdot) \neq \hat{\theta}(\cdot)$. Since their minmax value $\log \bar{C}_k$ is smaller than $\log \hat{C}_k$, why prefer the NML code? The real justification is the necessary condition in Theorem 4.1 and the maximum capacity estimator.

Because of the intuitive appeal of the ML estimator, it has been elevated by some statistician-philosophers to an ML principle in the hope of settling the problem of estimation. As we have seen, instead of the ML estimator we have established such a role for the maximum capacity estimator $\hat{\theta}(\cdot), \hat{k}(\cdot)$ as the MDL *principle* because of its provable optimality properties as well as the vast generalization that it covers even the estimation of the number of parameters and the structures. In fact, it cannot be dethroned without finding another more general and powerful yardstick than probability. To the credit of the ML estimator, it agrees with the maximum capacity estimator in the special case where the number of parameters is fixed, and its intuitively good properties have now been formally established.

In the spirit of the algorithmic complexity, Appendix A, we call

$$\ln 1/\hat{f}(\mathbf{x};\mathcal{M}) = \ln 1/\hat{f}(\mathbf{x};\hat{k}(\mathbf{x})) + \ln \hat{C} \qquad (4.55)$$

4.5 General and complete MDL principles

the *stochastic complexity*. Originally the name was used for the asymptotic version of the ideal code length of the non-optimized numerator [44]:

$$\ln 1/\hat{f}(\mathbf{x};k) = \ln 1/f(\mathbf{x};\hat{\theta}(\mathbf{x});k) + \ln \hat{C}_k \qquad (4.56)$$

$$\ln \hat{C}_k = \frac{k}{2}\ln\frac{n}{2\pi} + \ln\int |J(\theta)|^{1/2}d\theta + o(1), \qquad (4.57)$$

where $J(\theta)$ is the limit of the Fisher information matrix:

$$J_n(\theta) = \frac{1}{n}E_\theta\left\{\frac{\partial^2 \ln 1/f(X^n;\theta,k)}{\partial\theta_j\partial\theta_k}\right\} \to J(\theta). \qquad (4.58)$$

The expectation is with respect to the distribution $f(x^n;\theta,k)$.

The derivation of (4.57) is immediate under the CLT, especially if we assume the stronger point-wise convergence: the peak of $g(\hat{\theta};\theta,k)$, induced by $f(\mathbf{x};\theta,k)$ on the ML estimates, converges to

$$\max_{\hat{\theta}} g(\hat{\theta};\theta,k)/n^{k/2} \to \frac{|J(\theta)|^{1/2}}{(2\pi)^{k/2}}, \qquad (4.59)$$

and

$$\hat{C}_k = \left(\frac{n}{2\pi}\right)^{k/2}\int |J(\theta)|^{1/2}d\theta \times e^{o(1)}. \qquad (4.60)$$

Non-asymptotic calculation of the normalizing coefficient $\hat{C}_k$ in general is not easy. For discrete data the integral becomes a summation, for which in a number of cases, such as the multinomials, algorithms exist [22], [31]. Also for regression models built of logarithms of probabilities an algorithm exists, see [52].

The incomplete MDL principle is very general, because almost anything can be encoded and the basic requirement of just decodability is broad. The intent was that we consider a nested set of model classes such as $\mathcal{M}_k \subset \mathcal{M} \subset \cdots$ with the property that shorter and shorter code lengths are required to encode the supersets. To illustrate the process take an integer n as the object to be encoded without any family of distributions. As the first "model" take the set $\{1,2,\ldots,2^m\}$, where m is the smallest integer such that n belongs to the set, or that $\log_2 n \le m$. We need to encode the model, or the number m. We repeat the argument and get the next model, or the smallest integer k such that $\log_2 m \le k$. This process ends when the last model has only one element $1 = 2^0$. The total code length is about $L(n) = \log_2^*(n) = \log_2 n + \log_2\log_2 n + \cdots$, the sum ending with the last positive iterated logarithm value, derived in Appendix B.

Since the negative logarithm of a distribution $P(\theta, k)$ satisfies the Kraft inequality with equality, the general MDL principle is sometimes confused with the Bayesian method in the belief that a prior for the parameters is needed. This is not the case, nor is the entire Kraft inequality needed, which implies the enormous generality of the principle. The statement clearly requires only encoding of the given data sequence – never mind the others – for which we only need the decodable codeword of the estimated parameters $\hat{\theta}(\mathbf{x}), \hat{k}(\mathbf{x})$, and for this the entire code tree as the "prior" is not needed. As a simple example, consider the binary codeword 010010111. You can find the leaf it describes if you know only that 0 means the left son and 1 the right son, without knowing anything else about the tree. Moreover, 2 raised to the negative power of the length of this string is less than unity and defines a probability, without our having to know the distribution. This leads to tremendous simplification in applications. Suppose we need to encode a number of closed curves on a plane, each defining a different local model for the data, and we are interested in the number of the curves. This can be done by the popular "chain link" code, of which an example is a three-symbol alphabet, $-1,0,1$, where the meaning of the symbols is a step of a fixed length left at some agreed angle, a step in the same direction as the previous step, and a step to the right. This is a perfectly valid code, which can be decoded by knowing only the starting point and the string in the three symbols. A Bayesian would have to define an entire prior for the set of all closed curves, before the posterior for the number of the curves could be maximized.

The shortcoming of the general MDL principle is the fact that

$$F(\mathbf{x}) = f(\mathbf{x}; \hat{\theta}(\mathbf{x}), \hat{k}(\mathbf{x})) \times 2^{-L(\hat{\theta}(\mathbf{x}), \hat{k}(\mathbf{x}))} \tag{4.61}$$

is incomplete and hence non-optimal. This is relevant when the model class cannot be well defined, perhaps for the reason that we are hesitant about the nature of the statistical constraints in the data.

For those interested in data compression we conclude this section by deriving a complete Bayes type of mixture code different from (4.55):

$$f_w(\mathbf{x}) = \sum_k v(k) \int f(\mathbf{x}; \theta, k) w(\theta|k) \mathrm{d}\theta, \tag{4.62}$$

where $w(\theta|k)$ and $v(k)$ are priors. The mixture given by the integral could be maximized over k, which gives a criterion for order estimation. This is only a stop-gap solution to the estimation problem, for the result depends on the priors. It is true that if we write $w(\theta, k) = w(\theta|k)v(k)$ the dependence on the prior $w(\theta|k)$ vanishes as $n \to \infty$, because the integrand peaks at $f(\mathbf{x}; \hat{\theta}(\mathbf{x}), k)w(\hat{\theta}(\mathbf{x})|k)$ for each k, and becomes independent of the prior, although the result still depends on $v(k)$.

We nevertheless describe the derivation for k fixed as the solution to a minmax problem, whose originator I do not know, because it at least agrees with the optimal estimator asymptotically and, moreover, it serves as an introduction to the more relevant non-asymptotic minmax problems in the next chapter. For data compression the mixture code is fine, because often the data sequences are long and the code length for the maximizing $\hat{k}$ is ignorable.

Start with the familiar minmax problem in universal coding for $\mathcal{M}_k$:

$$\min_q \max_\theta D\left(f_{\theta,k} \| q\right), \tag{4.63}$$

where q ranges over all density functions, and

$$D(f_{\theta,k} \| q) = \int f_{\theta,k}(\mathbf{x}) \log \frac{f_{\theta,k}(\mathbf{x})}{q(\mathbf{x})} d\mathbf{x} \tag{4.64}$$

is the KL distance between the density function $f(\mathbf{x}; \theta, k)$, written as $f_{\theta,k}$, and q. This turns out to be very difficult to solve with no neat solution, and the problem was changed to

$$\max_w \min_q \int w(\theta) D(f_{\theta,k} \| q) d\theta. \tag{4.65}$$

This is more tractable since by the noise-free coding theorem, the minimizing q for all w is the mixture model

$$q_w(\mathbf{x}) = \int f(\mathbf{x}; \theta, k) w(\theta) d\theta. \tag{4.66}$$

To see this we have

$$\int w(\theta) d\theta \int f(\mathbf{x}; \theta, k) \ln \frac{f(\mathbf{x}; \theta, k)}{q(\mathbf{x})} d\mathbf{x}$$
$$= \int q_w(\mathbf{x}) \ln \frac{1}{q(\mathbf{x})} d\mathbf{x} + \int w(\theta) d\theta \int f(\mathbf{x}, \theta, k) \ln f(\mathbf{x}, \theta, k) d\mathbf{x}, \tag{4.67}$$

where in the first line we have switched the order of the two integrations. The second term on the second line does not depend on q, and by the noise-free coding theorem the minimizing q in the first term is $q_w(\mathbf{x})$.

Although the maximizing prior w^* is still difficult to find, the maxmin value, call it $\log C_{w^*}$, is formally Shannon's channel capacity. Asymptotically $\log C_{w^*}$, which can also be shown to agree with the maxmin value, is called the "regret," and the maximizing prior converges to Jeffrey's famous prior in Bayesian statistics.

There are several further studies of the general MDL principle in estimation but here we give just three: [14], [16], and [45].

5 Other optimality properties

There are other intuitively attractive properties we would like estimators to have. Such properties include consistency and certain minmax properties ensuring that an estimator works well even when the data are generated by the most difficult to estimate models. We show in this chapter that the ML estimator $\hat{\theta}(\cdot)$ and the maximum capacity estimator $\bar{\theta}(\cdot), \hat{k}(\cdot)$ satisfy such properties. These are in the same spirit as the celebrated Cramér–Rao inequality, [5], which sets a lower bound for the covariance of the estimated parameters. The two theorems below are, however, much stronger than the Cramér–Rao inequality, in which the equality is reached by the ML estimator only for a small subset of the models, and for all of them only asymptotically. Moreover, and even more importantly, the Cramér–Rao inequality actually leads to a dead end which blocks any foundation for an extensive theory, because there is no covariance inequality for the case where the number of parameters is also estimated.

5.1 Minmax problems

We consider the two minmax problems

$$\min_{\bar{\theta}(\cdot)} \max_{\theta,k} D(f_{\theta,k} \| \bar{f}_k) \tag{5.1}$$

and

$$\min_{\bar{\theta}(\cdot),\bar{k}(\cdot)} \max_{\theta,k} D(f_{\theta,k} \| \bar{f}), \tag{5.2}$$

where we have used the notation

$$f_{\theta,k} \quad \text{for} \quad f_{\theta,k}(\mathbf{y}) = f(\mathbf{y};\theta,k), \tag{5.3}$$

$$\bar{f}_k \quad \text{for} \quad \bar{f}_k(\mathbf{y}) = f(\mathbf{y};\bar{\theta}(\mathbf{y}),k)/\bar{C}_k, \tag{5.4}$$

$$\bar{f} \quad \text{for} \quad \bar{f}(\mathbf{y}) = \frac{f(\mathbf{y};\bar{\theta}(\mathbf{y}),\bar{k}(\mathbf{y}))/\bar{C}_{\bar{k}(\mathbf{y})}}{\bar{C}}. \tag{5.5}$$

These differ from the familiar minmax problem for universal data compression, (4.65), whose solution is the Bayes mixture with prior $\hat{w}(\theta)$ that gives Shannon's channel capacity as the "regret." These minmax problems are more appropriate for estimation, because the maximization is over parameters, for which the "regret" could not be solved, and one had to settle for maximization over priors. After all, the Bayes mixture with any prior works well only because the mixture includes the ML estimator, which asymptotically dominates. Otherwise a mixture is useless even for data compression.

Theorem 5.1 *Let $f(\mathbf{x}; \theta, k)$ be a continuous function of θ at all θ and $\mathbf{x}$. Then for all k and n the solution to the minmax problem*

$$\min_{\bar{\theta}(\cdot)} \max_{\theta \in \Omega} D(f_{\theta,k} \| \bar{f}_k) \tag{5.6}$$

is $\hat{\theta}(\cdot)$.

Proof To simplify the notation we drop the index k, which is held constant, and we write for instance $\bar{f}$ for $\bar{f}_k$. We emphasize that we are proving the first minmax theorem where k is fixed without showing the dependence of the various density functions like f_k on k. Let

$$\Delta(\bar{f}, \theta) = D(f_\theta \| \bar{f}) - D(f_\theta \| \hat{f}) = \int f_\theta(\mathbf{y}) \ln \frac{\hat{f}(\mathbf{y})}{\bar{f}(\mathbf{y})} d\mathbf{y}, \tag{5.7}$$

and let $\bar{A} = \{\mathbf{y} : \bar{\theta}(\mathbf{y}) \neq \hat{\theta}(\mathbf{y})\}$. For discrete data the integral is a sum. Clearly for all $\bar{f}$

$$\max_{\theta \in \Omega} \Delta(\bar{f}, \theta) \geq \sup_{\theta = \bar{\theta}(\mathbf{x}): \mathbf{x} \in \bar{A}} \Delta(\bar{f}, \theta). \tag{5.8}$$

Also

$$\min_{\bar{f}} \max_{\theta \in \Omega} \Delta(\bar{f}, \theta) \geq \inf_{\bar{f}} \sup_{\theta = \bar{\theta}(\mathbf{x}): \mathbf{x} \in \bar{A}} \Delta(\bar{f}, \theta). \tag{5.9}$$

We need to use inf and sup, because min and max do not exist for $\mathbf{x} \in \bar{A}$.

We prove in the following lemma that the inequality

$$\sup_{\theta = \bar{\theta}(\mathbf{x}): \mathbf{x} \in \bar{A}} \Delta(\bar{f}, \theta) \geq 0 \tag{5.10}$$

holds for all $\bar{\theta}(\cdot)$ at $\mathbf{x} \in \bar{A}$ and $\bar{f}(\cdot)$ it determines, which then implies the theorem, because $\Delta(\hat{f}, \theta) = 0$ for all θ.

Lemma 5.1 *Let the conditions for $f(\mathbf{x};\theta)$ in Theorem 5.1 hold. Then (5.10) holds.*

Proof The idea of the proof is to argue that for each $\mathbf{x} \in \bar{A}$ and parameter estimator $\bar{\theta}(\cdot)$ with the distribution that it defines $\bar{f}(\cdot)$ we can find a better distribution, say $\tilde{f}(\cdot)$, in the sense that $\Delta(\bar{f},\hat{\theta}(\mathbf{x})) > \Delta(\tilde{f},\hat{\theta}(\mathbf{x}))$, if the negation of (5.10)

$$\Delta(\bar{f},\hat{\theta}(\mathbf{x})) \leq -c, \tag{5.11}$$

holds for positive c. These two inequalities imply that also $\Delta(\tilde{f},\hat{\theta}(\mathbf{x})) < -c$. Repeating this until the limit of the better distributions gives $\hat{f}(\mathbf{x})$, and we remove point $\mathbf{x}$ from $\bar{A}$. This is continued until $\bar{A}$, where we need to prove (5.10), is empty.

Arguing then by way of contradiction, assume that

$$\sup_{\theta=\bar{\theta}(\mathbf{x}):\mathbf{x}\in\bar{A}} \Delta(\bar{f},\theta) = -c \tag{5.12}$$

holds for some positive c. For all $\bar{f}$ and $\mathbf{x} \in \bar{A}$ (5.11) holds because the left hand side cannot exceed the supremum.

We show first that for all $\mathbf{x}$ and $\bar{f}$ such that $\bar{\theta}(\mathbf{x}) \neq \hat{\theta}(\mathbf{x})$ there is another better $\tilde{f}$ for which

$$D(f_{\hat{\theta}(\mathbf{x})} \| \bar{f}) > D(f_{\hat{\theta}(\mathbf{x})} \| \tilde{f}), \tag{5.13}$$

and for which $\tilde{C} > \bar{C}$.

We prove the lemma first for discrete data and $\bar{C} = \bar{C}_k > 1$, in which case all the density functions are probabilities no greater than unity, and then the case where $\bar{C} \leq 1$. Finally, we explain that it holds even for continuous data. Consider an estimator function $\tilde{\theta}(\cdot)$ and the distribution $\tilde{f}$ it determines as follows:

$$f(\mathbf{x};\tilde{\theta}(\mathbf{x})) = f(\mathbf{x};\bar{\theta}(\mathbf{x})) + \delta \leq f(\mathbf{x};\hat{\theta}(\mathbf{x})), \tag{5.14}$$

while $\tilde{\theta}(\mathbf{y}) = \bar{\theta}(\mathbf{y})$ for $\mathbf{y} \neq \mathbf{x}$. Such a $\tilde{\theta}(\mathbf{x})$ exists by the continuity of $f(\mathbf{x};\theta)$ as a function of θ. Here δ is positive, with its size to be selected. Then

$$\tilde{f}(\mathbf{x}) = \frac{f(\mathbf{x};\tilde{\theta}(\mathbf{x}))}{\tilde{C}} = \frac{f(\mathbf{x};\bar{\theta}(\mathbf{x})) + \delta}{\bar{C} + \delta} = \bar{f}(\mathbf{x}) \frac{1 + \delta/f(\mathbf{x};\bar{\theta}(\mathbf{x}))}{1 + \delta/\bar{C}}, \tag{5.15}$$

$$\tilde{f}(\mathbf{y}) = \frac{f(\mathbf{y};\bar{\theta}(\mathbf{y}))}{\bar{C} + \delta} = \bar{f}(\mathbf{y}) \frac{1}{1 + \delta/\bar{C}} \text{ for } \mathbf{y} \neq \mathbf{x}. \tag{5.16}$$

We keep $\mathbf{x}$ fixed and write $\bar{\theta}(\mathbf{x}) = \bar{\theta}_{\mathbf{x}}$ and $\hat{\theta}(\mathbf{x}) = \hat{\theta}_{\mathbf{x}}$. We have

$$\Delta(\bar{f}, \hat{\theta}_{\mathbf{x}}) - \Delta(\tilde{f}, \hat{\theta}_{\mathbf{x}}) = \sum_{\mathbf{y}} f(\mathbf{y}; \hat{\theta}_{\mathbf{x}}) \log \frac{\tilde{f}(\mathbf{y})}{\bar{f}(\mathbf{y})}. \tag{5.17}$$

By summing over the data $\mathbf{y}$ first over $\mathbf{y} \neq \mathbf{x}$ and then adding the single term at $\mathbf{x}$ we get

$$\Delta(\bar{f}, \hat{\theta}_{\mathbf{x}}) - \Delta(\tilde{f}, \hat{\theta}_{\mathbf{x}}) = -\log(1 + \delta/\bar{C}) + f(\mathbf{x}; \hat{\theta}_{\mathbf{x}}) \log(1 + \delta/f(\mathbf{x}; \bar{\theta}_{\mathbf{x}})). \tag{5.18}$$

It is positive if

$$f(\mathbf{x}; \hat{\theta}_{\mathbf{x}}) \log(1 + \delta/f(\mathbf{x}; \bar{\theta}_{\mathbf{x}})) \geq \delta/\bar{C}, \tag{5.19}$$

since $\log(1 + \delta/\bar{C}) \leq \delta/\bar{C}$. With the notation $\bar{\delta} = \delta/f(\mathbf{x}; \bar{\theta}_{\mathbf{x}})$ and for $\bar{\delta} < 1$ expand $\log(1 + \bar{\delta})$ into a Taylor series, which by the alternating series theorem gives

$$\log(1 + \delta/f(\mathbf{x}; \bar{\theta}_{\mathbf{x}})) > \bar{\delta} - \bar{\delta}^2/2. \tag{5.20}$$

With this, (5.19) is true if

$$f(\mathbf{x}; \hat{\theta}_{\mathbf{x}}) \bar{\delta} (1 - \bar{\delta}/2) \geq f(\mathbf{x}; \bar{\theta}_{\mathbf{x}})(\bar{\delta}/C). \tag{5.21}$$

This can be achieved if we pick δ to satisfy

$$0 < \delta \leq 2 f(\mathbf{x}; \bar{\theta}_{\mathbf{x}}) / \left(1 - \frac{f(\mathbf{x}; \bar{\theta}_{\mathbf{x}})/f(\mathbf{x}; \hat{\theta}_{\mathbf{x}})}{\bar{C}}\right), \tag{5.22}$$

and we have shown (5.13). We see that the bigger the initially selected $f(\mathbf{x}; \bar{\theta}_{\mathbf{x}})$ is, the bigger δ we may select and still guarantee condition (5.19) to hold.

By repeating this for the same $\mathbf{x}$, we arrive at a sequence of estimators $\bar{\theta}(\cdot), \tilde{\theta}(\cdot), \ldots, \hat{\theta}(\cdot)$, and the distributions they define such that the last one evaluated equals $\hat{f}(\mathbf{x})$ at $\mathbf{x}$. Since the sequence of the increments δ does not shrink, there will be only a finite number of steps to reach $\hat{f}(\mathbf{x})$, and we reduce $\bar{A}$ by the point $\mathbf{x}$. This reduces the right hand side of (5.9), and we now need to prove the lemma for a smaller set $\bar{A}$. Continuing in this way we reduce $\bar{A}$ to a one-element set, for which the limit of the estimators defines $\hat{f}(\cdot)$, for which $\Delta(\hat{f}, \theta) = 0$, and this proves the lemma.

We then assume $\bar{C} \leq 1$. This time define

$$\tilde{f}(\mathbf{x}) = \frac{f(\mathbf{x};\tilde{\theta}(\mathbf{x}))}{\tilde{C}} = \frac{f(\mathbf{x};\bar{\theta}(\mathbf{x})) + \delta_\mathbf{x}}{\bar{C} + \epsilon} = \bar{f}(\mathbf{x})\frac{1 + \delta_\mathbf{x}/f(\mathbf{x};\bar{\theta}(\mathbf{x}))}{1 + \epsilon/\bar{C}}, \quad (5.23)$$

$$\tilde{f}(\mathbf{y}) = \bar{f}(\mathbf{y})\frac{1 - \delta_\mathbf{y}/f(\mathbf{y};\bar{\theta}(\mathbf{y}))}{1 + \epsilon/\bar{C}} \quad \text{for } \mathbf{y} \neq \mathbf{x}, \quad (5.24)$$

where ϵ and $\{\delta_\mathbf{y}\}$ are positive satisfying $\epsilon = \delta_\mathbf{x} - \sum_{\mathbf{y}\neq\mathbf{x}} \delta_\mathbf{y} > 0$. We shall see that this is possible. We can even pick $\delta_\mathbf{y}$ to be constant and just a fraction of $\delta_\mathbf{x}$. However, we do not need to be specific.

We have

$$\Delta(\bar{f}, \hat{\theta}_\mathbf{x}) - \Delta(\tilde{f}, \hat{\theta}_\mathbf{x}) = f(\mathbf{x};\hat{\theta}_\mathbf{x})\log(1 + \delta_\mathbf{x}/f(\mathbf{x};\bar{\theta}_\mathbf{x})) - \log(1 + \epsilon/\bar{C})$$
$$- \sum_{\mathbf{y}\neq\mathbf{x}} f(\mathbf{y};\hat{\theta}_\mathbf{x})\log(1 - \delta_\mathbf{y}/f(\mathbf{y};\bar{\theta}_\mathbf{x})). \quad (5.25)$$

Since the last term is positive the sufficient condition for (5.25) to be positive is just (5.19) with δ replaced by $\delta_\mathbf{x}$ and ϵ thus

$$f(\mathbf{x};\hat{\theta}_\mathbf{x})\log(1 + \delta_\mathbf{x}/f(\mathbf{x};\bar{\theta}_\mathbf{x})) \geq \epsilon/\bar{C}. \quad (5.26)$$

If, moreover, we write $\bar{\delta}_\mathbf{x} = \delta_\mathbf{x}/f(\mathbf{x};\bar{\theta}_\mathbf{x})$ then for $\epsilon = \bar{\delta}_\mathbf{x}\bar{C}/2$ (5.26) holds if $0 < \delta_\mathbf{x} \leq f(\mathbf{y};\bar{\theta}(\mathbf{x}))$. Then also $\tilde{C} = \bar{C} + \epsilon = \bar{C}(1 + \bar{\delta}_\mathbf{x}/2)$, and hence is greater than $\bar{C}$. Continuing in this manner we get an estimator $\bar{\theta}(\cdot)$ and some $\bar{f}$ with $\bar{A}$ it defines for which $\bar{C} > 1$, and hence the lemma holds even when the initial $\bar{C} \leq 1$.

For real-valued data we need to quantize the data to obtain probability functions from the density functions as well as to define $\tilde{\theta}$ and $\tilde{f}$ such that they differ from $\bar{\theta}$ and $\bar{f}$, respectively, in a small neighborhood of $\mathbf{x}$. Since Theorem 5.1 holds for all such quantizations including their limits, it holds for real-valued data as well.

If for some discrete data $\mathbf{x}$, $f(\mathbf{x};\hat{\theta}(\mathbf{x}),k) = 1$, the minmax value is $\log\hat{C}_k$. Examples of special data $\mathbf{x}$ such that the minmax value is $\log\hat{C}$ exist for Markov chains. For the Bernoulli class the special sequences consist of all 0s or all 1s.

Theorem 5.2 *Let $f(\mathbf{x};\theta,k)$ be a continuous function of θ at all θ and $\mathbf{x}$. Then for all k and n the solution to the minmax problem (5.2) is $\hat{\theta}(\cdot)$ and $\hat{k}(\cdot)$.*

Proof The proof is virtually the same as the proof of Theorem 5.1 and we describe only the main steps for $\bar{C} > 1$. The key difference is that we consider the complete pdfs $\bar{f}(\mathbf{y}) = f(\mathbf{y}; \bar{k}(\mathbf{y}))/\bar{C}$.

Let $\bar{A} = \{\mathbf{x} : (\bar{\theta}(\mathbf{x}), \bar{k}(\mathbf{x})) \neq (\hat{\theta}(\mathbf{x}), \hat{k}(\mathbf{x}))\}$. We keep $\mathbf{x}$ fixed and write $(\hat{\theta}, \hat{k}) = (\hat{\theta}(\mathbf{x}), \hat{k}(\mathbf{x}))$. Consider an estimator $\bar{\theta}(\cdot), \bar{k}(\cdot) \neq \hat{\theta}(\cdot), \hat{k}(\cdot)$ and the distribution for discrete data it defines, $\bar{f}(\mathbf{y}) = f(\mathbf{y}; \bar{k}(\mathbf{y}))/\bar{C}$. We show that there is a better one $\tilde{f}$:

$$D(f_{\hat{\theta},\hat{k}} \| \bar{f}) > D(f_{\hat{\theta},\hat{k}} \| \tilde{f}), \tag{5.27}$$

for which $\tilde{C} > \bar{C}$.

Let $(\tilde{\theta}(\mathbf{y}), \tilde{k}(\mathbf{y}))$ be another estimator and the density function $\tilde{f}$ it defines be as follows:

$$\text{for } \mathbf{x} \in \bar{A}, \quad \tilde{f}(x) = \frac{\bar{f}(\mathbf{x}; \bar{k}(\mathbf{x})) + \delta}{\bar{C} + \delta} = \bar{f}(\mathbf{x}) \frac{1 + \delta/\bar{f}(\mathbf{x}; \bar{k}(\mathbf{x}))}{1 + \delta/\bar{C}} \tag{5.28}$$

$$\text{else, } \tilde{f}(\mathbf{y}) = \frac{\bar{f}(\mathbf{y}; \bar{k}(\mathbf{y}))}{\bar{C} + \delta} = \bar{f}(\mathbf{y}) \frac{1}{1 + \delta/\bar{C}}. \tag{5.29}$$

We have

$$\Delta(\bar{f}, (\hat{\theta}, \hat{k})) - \Delta(\tilde{f}, (\hat{\theta}, \hat{k})) = \sum_{\mathbf{y}} f(\mathbf{y}; \hat{\theta}, \hat{k}) \log \frac{\tilde{f}(\mathbf{y})}{\bar{f}(\mathbf{y})} \tag{5.30}$$

$$= -\log(1 + \delta/\bar{C}) + f(\mathbf{x}; \hat{\theta}, \hat{k}) \log(1 + \delta/\bar{f}(\mathbf{x}; \bar{k}(\mathbf{x}))) \tag{5.31}$$

$$\geq -\delta/\bar{C} + f(\mathbf{x}; \hat{\theta}, \hat{k})) \log(1 + \delta/\bar{f}(\mathbf{x}; \bar{k}(\mathbf{x}))). \tag{5.32}$$

Putting $\bar{\delta} = \delta/\bar{f}(\mathbf{x}; \bar{k}(\mathbf{x}))$ and expanding the logarithm into a Taylor series we get the sufficient condition

$$1 - 1/\bar{C} \geq \frac{\delta}{2\bar{f}(\mathbf{x}; \bar{k}(\mathbf{x}))} \tag{5.33}$$

for the inequality

$$\Delta(\bar{f}, (\hat{\theta}, \hat{k})) - \delta(\tilde{f}, (\hat{\theta}, \hat{k})) \geq 0, \tag{5.34}$$

which holds if δ is small enough to satisfy

$$0 < \delta \leq 2f(\mathbf{x}; \bar{\theta}_{\mathbf{x}}, \bar{k}(\mathbf{x})) \left(1 - \frac{f(\mathbf{x}; \bar{\theta}_{\mathbf{x}}, \bar{k}(\mathbf{x}))/f(\mathbf{x}; \hat{\theta}_{\mathbf{x}}, \hat{k}(\mathbf{x}))}{\bar{C}}\right). \tag{5.35}$$

The same arguments as in the proof of Theorem 5.1 finish the proof.

5.2 Consistency

The consistency in probability of the ML estimates of the real-valued parameters

$$\int f(x^n; \theta, k)(\|\hat{\theta}(x^n) - \theta\| > c/\sqrt{n})\, dx^n < \epsilon \qquad (5.36)$$

for all ϵ and c when the data length n exceeds some number is an old result, which follows from the CLT theorem. Moreover, the Cramér–Rao inequality establishes an in effect optimal rate of convergence for the ML estimator in terms of the covariance.

The consistency of the estimates of the number of real-valued parameters has been shown for many criteria in the almost sure sense; see, for instance, [7] and [15]. The problem is more intricate, and the appropriate measure depends on the two main types of models, the "batch" models considered hitherto in this book and the sequential models considered in Chapter 9, which define random processes.

If we model the physical machinery by a distribution $f(y^n; \hat{\theta}(x^n), \hat{k}(x^n))$, we would like the estimates $\hat{\theta}(\cdot), \hat{k}(\cdot)$ from repeated data of the same length to be close to each other; the closer, the longer the data we consider. This can be analyzed meaningfully as consistency in probability, or in terms of the KL distance $D(f_{\theta,k} \| f_{\hat{\theta}(\cdot), \hat{k}(\cdot)})$, but not in terms of almost sure behavior on an infinite sequence, which is appropriate and natural only for the random process models. These two types of consistencies and fitted models should not be confused, for if they are strange and completely misleading results about the behavior of estimates and criteria are obtained.

It was shown in [7] that by removing an upper bound for the order estimates the criterion

$$\min_k \log 1/f(x^n; \hat{\theta}(x^n), k) + (k/2) \log n, \qquad (5.37)$$

called BIC, makes the estimates of the number of parameters consistent almost surely for a uniform independent identically distributed (iid) process. Quite surprisingly it was shown that the estimates obtained with a number of other criteria, including the almost equivalent looking NML criterion

$$\min_k \log 1/\hat{f}(x^n; k) = \min_k \log 1/f(x^n; \hat{\theta}(x^n), k) + (k/2) \log(n/(2\pi)) + o(1) \qquad (5.38)$$

are not consistent almost surely for the same iid process. This criterion is defined by a batch model, which for large n is essentially the complete MDL criterion, and the almost sure measure for consistency is inappropriate. The BIC criterion has neither information nor a probability theoretic interpretation, and it does not matter which measure for consistency is selected.

If the consistency is measured in the almost sure sense as in [7] we must not compare BIC with the NML criterion, which is equivalent with the negative logarithm of a batch model, and a collection of batch models of varying quantities of data does not give a random process model. Rather, it should be compared with a criterion appropriate for random process models, such as the finite order Markov chains studied in [7], of which the uniform iid process is a special case. Each chain has a definite finite order which cannot change along the string. The model that satisfies the necessary conditions for optimality of our probability maximization postulate gives the sequential normalized maximum likelihood (SNML) criterion, (9.2) and (9.4),

$$\min_k \sum_{t=\tau(k)}^{n} \log 1/f(x_t|x^{t-1}; \hat{\theta}(x^t), k) + \sum_t \log K(x^t), \quad (5.39)$$

where the maximized likelihood in batch models is replaced by the larger sequentially maximized likelihood, defined by the ML estimates and calculated from the prefixes x^t of the string x^n, starting at the smallest point $\tau(k)$, where k parameters can be solved. The second sum, defined by the normalizing coefficient, is the maximum sequential capacity.

We show now that for the family of Markov chains, or even for the variable order chains, which are beyond the capability of BIC, the iid SNML model

$$\log 1/\hat{P}_{S1}(x^n) = \log \binom{n}{n_1} + \log(n+1) \quad (5.40)$$

$$= nh(n_0/n) + \frac{1}{2}\log n + o(1), \quad (5.41)$$

defined by the Laplace estimator, (9.28), is the minimizing random process model that gives almost sure consistency for typical data generated by the Bernoulli process $P(0) = p_0$, including the uniform $p_0 = 1/2$. Write $p_1 = 1 - p_0$ and let n_i be the number of times the symbol i occurs in the string x^n. Here $h(p)$ is the binary entropy function, and the second equality holds provided n_0/n is bounded away from 0 and 1. Of the other Markov chain models consider the first order

Markov chain

$$\log 1/\hat{P}_{S2}(x^n) = n_0 h(n_{0|0}/n_0) + n_1 h(n_{0|1}/n_1) + \frac{1}{2}[\log n_0 + \log n_1] + o(1), \tag{5.42}$$

where n_i is the number of times the state i, $i = 0, 1$, occurs in the string x^n, and $n_{j|i}$ denotes the number of these occurrences for which the "next" symbol is j.

The code length criterion (5.41) clearly cannot be shortest for all strings and hence give consistency. Rather, the appropriate measure of the goodness of a criterion is almost sure consistency for it means that the probability of the set of infinite strings where it gives consistency is 1. The probability measure for sets of infinite strings is the unique Kolmogorov extension of the Bernoulli measure.

Define the set of typical strings for data generated by the Bernoulli model as in [4]:

$$A_\epsilon(n) = \{x^n : \max_i |n_i/n - p_i| \leq \epsilon\}, \tag{5.43}$$

for all ϵ, however small, and large enough n. This is clearly equivalent to

$$B_\epsilon(n) = \{x^n : D(s(\mathbf{x}^n)\|p) \leq \epsilon\}, \tag{5.44}$$

not necessarily for the same ϵ, where $s = (n_0/n, n_1/n)$ and $p = (p_0, p_1)$. It was shown by the theory of types in [4], see also [6], that

$$\sum_n Pr\{x^n : D(s(x^n)\|p) > \epsilon\} < \infty. \tag{5.45}$$

In fact,

$$\log 1/Pr(x^n; p) = \sum_i n_i \log 1/p_i = n(h(s_0) + D(s\|p)) \geq n D(s\|p). \tag{5.46}$$

Hence,

$$Pr\{x^n : D(s(x^n)\|p) > \epsilon\} \leq |B_\epsilon(n)| 2^{-n\epsilon} \leq (n+1)^2 2^{-n\epsilon}, \tag{5.47}$$

which is summable.

The same holds for the sets $\bar{A}_\epsilon(n)$ of atypical strings, i.e. strings that are non-typical. To see what this means consider the sequence of events $\bar{A}_\epsilon(1), \bar{A}_\epsilon(2), \ldots$, among which infinitely many "occur," i.e. are non-empty. This can be written as the set

$$\limsup_{n \to \infty} \bar{A}_\epsilon(n) = \bigcap_{m=1}^{\infty} \bigcup_{n=m}^{\infty} \bar{A}_\epsilon(n). \qquad (5.48)$$

By the Borel–Cantelli lemma the probability of this set is 0, and hence the probability of the complement set is 1. Since this holds for all ϵ, it must be the case that for each typical string $n_i/n = p_i$ for n greater than some number, say $N(x^n)$, that depends on the string.

Compare the two code lengths (5.41) and (5.42). For $n \geq N(x^n)$, $h(n_0/n) = h(p_0)$ equals the conditional entropy $(n_0/n)h(n_{0|0}/n_0) + (n_1/n)h(n_{0|1}/n_1) = p_0 h(p_0) + p_1 h(p_1)$. In addition, $\frac{1}{2} \log n < \frac{1}{2}[\log(np_0) + \log(np_1)]$. Hence $\hat{P}_{S1}(x^n) > \hat{P}_{S2}(x^n)$ for $n \geq N(x^n)$. The same clearly holds also for all higher order Markov chains as well, and the maximum capacity estimates without any bound on the number of parameters are consistent almost surely.

This resolves the puzzle in [7], which suggests that there are anomalous models which only the BIC could estimate consistently. We see that this is not the case and that there is no need at all for ad hoc criteria.

We show in the rest of this section that the estimates of the batch models satisfying our probability maximization postulate are not only consistent on the relevant finite data in the sense of the KL distance but also the consistency rate is the fastest possible. Actually, such results were proven in [41], [42], but since at the time the present theory of estimation was not in existence the results were aimed at coding without the realization of their applicability to order estimation.

We consider the class of parametric pdfs $\mathcal{M}_k = \{f(x^n; \theta, k)\}$, where the parameter vector $\theta = \theta_1, \ldots, \theta_k$ ranges over a bounded subset $\Omega = \Omega^k$ of the k-dimensional euclidian space. The parameters are taken to be "free" in the sense that distinct values for θ specify distinct probability measures.

Theorem 5.3 *Assume that the ML estimates $\hat{\theta}(x^n)$ satisfy the consistency property (5.36) in probability in the interior of Ω. If $q(x^n)$ is a density function, in particular $\bar{f}(x^n)$ defined by an estimator $(\bar{\theta}(x^n), \bar{k}(x^n))$, then for all positive numbers ϵ, all sufficiently large n, and all $\theta \in \Omega^k$, except in a set $A_{n,\theta}$ whose volume goes to zero as $n \to \infty$,*

$$E_{\theta,k} \log \frac{f(x^n; \theta, k)}{q(x^n)} \geq \frac{k - \epsilon}{2} \log n. \qquad (5.49)$$

The mean is taken relative to $f_{\theta,k}(x^n) = f(x^n; \theta, k)$. Under the added requirement $\hat{C} = o(\log n)$ the fastest rate by which $D(f_{\theta,k}\|q)/n$ converges to zero is obtained for $q(x^n) = \hat{f}(x^n)$.

We give the original proof, which has somewhat intricate steps, but it is intuitive. There is a different proof of an extension of the theorem in [30], which although shorter is also quite intricate. The idea of the proof we give is based on the obvious fact that if any single density function is to be close to several data generating density functions in the sense of the KL distance, it will have to distribute its probability mass so that none is favored, and each is allocated its share. If the ML estimator satisfies the CLT, which is true for most model classes satisfying the conditions stated, then we know the amount of the probability mass it can distribute around each point θ. Well, there is only so much probability mass available for any density function, and there is the limit which cannot be beaten except for models in a small set.

The strength of the theorem is that nothing specific is needed about the density function $q(x^n)$ such as it must not depend on θ. If it does, it will fall in the exceptional set, and we do not need to put such a condition for $q(x^n)$. This becomes relevant in prediction, where a predictor which depends on the process we want to predict can do better than the lower bound in the theorem. However, such a case falls in the exceptional set without which the theorem would not hold.

Proof Let $C_{n,c}(\theta)$ be a hypercube of edge length $\Delta_n = c/\sqrt{n}$, centered at θ, where c is a constant to be selected, and let $X_n(\theta) = \{x^n | \hat{\theta}(x^n) \in C_{n,c}(\theta)\}$. Denote the probability of this set under the distribution $f(x^n; \theta, k)$ by $P_n(\theta)$. Because of the assumed consistency of the estimator, $P_n(\theta)$ satisfies the inequality

$$P_n(\theta) \geq 1 - \delta(c) \qquad (5.50)$$

for all n greater than some number, say n_c, that depends on c. Moreover, $\delta(c)$ can be made as small as we please by selecting c and $C_{n,c}(\theta)$ large enough.

Consider next a density function $q(x^n)$, and let $Q_n(\theta)$ denote the probability mass it assigns to the set $X_n(\theta)$. The ratio $f(x^n; \theta, k)/P_n(\theta)$ defines a distribution on $X_n(\theta)$, as does $q(x^n)/Q_n(\theta)$. By the noiseless coding theorem, applied to these two distributions, we get

$$\int_{X_n(\theta)} f(x^n;\theta,\gamma) \log \frac{f(x^n;\theta,k)}{q(x^n)} dx^n \geq P_n(\theta) \log \frac{P_n(\theta)}{Q_n(\theta)}. \quad (5.51)$$

For a positive number ϵ, $\epsilon < 1$, let $A_\epsilon(n)$ be the subset of Ω^k such that the left hand side of (5.51), denoted $T_n(\theta)$, satisfies the inequality at each θ

$$T_n(\theta) < (1-\epsilon) \log n^{k/2} \quad (5.52)$$

and hence puts too much probability mass around each θ. From (5.51) and (5.52) we get

$$-\log Q_n(\theta) < \left[\frac{(1-\epsilon)}{P_n(\theta)} - \frac{\log P_n(\theta)}{\log n^{k/2}}\right] \log n^{k/2}, \quad (5.53)$$

which holds for $\theta \in A_\epsilon(n)$. Replace $P_n(\theta)$ by its lower bound $1-\delta(c)$ in (5.50), which does not reduce the right hand side. Pick c so large that $\delta(c) \leq \epsilon/2$ for all n greater than some number n_ϵ. The first term within the square bracket is then strictly less than unity, and since the second term is bounded from above by $(-\log(1-\epsilon/2))/\log n^{k/2}$ and hence converges to zero with growing n, the expression within the brackets is less than some α such that $0 < \alpha < 1$, for all sufficiently large n, say larger than some n'_ϵ. Therefore,

$$Q_n(\theta) > n^{-\alpha k/2} \quad (5.54)$$

for $\theta \in A_\epsilon(n)$ and n larger than n'_ϵ. The plan is to show that the volume of the set $A_\epsilon(n)$ cannot be large. This is the hardest part in the proof, and we need a partition of the set Ω^k into the hypercubes $C_{n,c}(\theta_i)$. Let $\Omega_{n,c} = \{\theta_1, \theta_2, \ldots\}$ be the centers of the hypercubes, say $m_{n,c}$ of them. The sets $X_{n,c}(\theta_i)$ define a corresponding partition of the set of all sequences of length n.

Let $B_\epsilon(n)$ be the smallest set of the centers of the hypercubes which cover $A_\epsilon(n)$, and let ν_n be the number of the elements in $B_\epsilon(n)$. Then the volume V_n of $A_\epsilon(n)$ is bounded by the total volume of the ν_n hypercubes, or

$$V_n \leq \nu_n c^k / n^{k/2}. \quad (5.55)$$

From (5.54) and the fact that the sets $X_n(\theta_i)$ are disjoint we get further

$$1 \geq \sum_{\theta_i \in B_\epsilon(n)} Q_n(\theta_i) \geq \nu_n n^{-\alpha k/2}, \quad (5.56)$$

which gives an upper bound for ν_n. From (5.55) we then get the desired inequality

$$V_n \leq c^k n^{(\alpha-1)k/2}, \tag{5.57}$$

which shows that $V_n \to 0$ as n grows to infinity.

Using the inequality $\ln z \geq 1 - 1/z$ for $z = f(x^n;\theta,k)/q(x)$ we get

$$\int_{\bar{X}_n(\theta)} f(x^n;\theta,k) \ln[f(x^n;\theta,k)/q(x^n)] dx^n \geq Q_n(\theta) - P_n(\theta) > -1, \tag{5.58}$$

where $\bar{X}$ denotes the complement of X. To finish the proof of the first part let $\theta \in \Omega^k - A_\epsilon(n)$. Then the opposite inequality, $\geq$, in (5.52) holds. By adding the left hand sides of (5.51) and (5.58) we get

$$E_\theta \ln[f(x^n;\theta,k)/q(x^n)] > (1-\epsilon) \ln n^{k/2} - 1, \tag{5.59}$$

which concludes the proof of (5.49).

To prove the second part of the theorem we have from (4.55) and (4.57)

$$D(f_{\theta,k} \| \hat{f}) = \int f_{\theta,k}(y^n) \log \frac{f_{\theta,k}(y^n)}{f(y^n;\hat{\theta}(y^n),\hat{k}(y^n))} d(y^n) + \log \hat{C}_{\hat{k}} + \log \hat{C}, \tag{5.60}$$

where we wrote $\hat{k} = \hat{k}(y^n)$. The first term on the right hand side is non-positive since

$$f(y^n;\theta,k) \leq f(y^n;\hat{\theta}(y^n),k) \tag{5.61}$$

for all k, including $k = \hat{k}$. We then have the inequalities

$$\frac{(k-\epsilon)}{2} \log n \leq D(f_{\theta,k} \| \hat{f}) \leq \log \hat{C}_{\hat{k}} + o(\log n). \tag{5.62}$$

Dividing by n we see that $\hat{k}(x^n) \to k$, and noting that the first inequality holds for all ϵ the proof follows from (4.60).

6 Interval estimation

Now that we know how to estimate real-valued parameters optimally the question arises of how confident we can be in the estimated result, which requires infinite precision numbers. It is clear that if we repeat the estimation on a new set of data, generated by the same physical machinery, the result will not be the same. It seems that the model class is too rich for the amount of data we have. After all, if we fit Bernoulli models to a binary string of length n, there cannot be more than 2^n properties in the data that we can learn, even if no two strings have a common property. And yet the model class has a continuum of parameter values, each representing a property.

One way to balance the learnable information in the data and the model class is to restrict the precision in the estimated parameters. However, we should not take any fixed precision, for example each parameter quantized to two decimals, because that does not take into account the fact that the sensitivity of models with respect to changes in the parameters depends on the parameters. The problem is related to statistical robustness, which while perfectly meaningful is based on practical considerations such as a model's sensitivity to outliers rather than on any reasonably comprehensive theory. If we stick to the model classes of interest in this book the parameter precision amounts to interval estimation. Here we are facing a difficulty: how can we talk about optimal estimation of intervals? It seems that we cannot talk about necessary conditions for optimality without an ML estimator, since the interval consisting of the entire parameter space encloses the probability mass unity. This is a serious difficulty which has plagued hypothesis testing and has never been resolved in a satisfactory manner to allow us to talk about optimal confidence intervals and optimal confidence in test results. Traditionally, confidence questions have been mired in the meaningless problem of how to avoid assigning probabilities to the unique non-random "truth."

We plan to tackle the problem differently. We want to find out how well we can separate the winning infinite precision maximum likelihood model from the nearby similarly optimal models. We want to do it for any number of data and even locally. This amounts to being able to define ML estimates of intervals, which then allows us to obtain optimal interval estimators in the same sense as the ML estimator does for points, i.e. they satisfy necessary conditions for optimality at all data. They also define the maximum capacity for intervals, which is strictly smaller than the capacity $\log \hat{C}_k$ for point estimates.

6.1 Optimum intervals

We first study the case where instead of the ML estimate $\hat{\theta}(\mathbf{x})$ an interval which includes it is estimated. In the next section we then extend the estimator to a maximum capacity estimator. If we wish to estimate an interval in which the ML estimate $\hat{\theta}(\mathbf{x})$ falls it is clear that all ML estimates $\hat{\theta}(\mathbf{y})$ that fall in the same interval will have to be regarded as equivalent. The natural requirement appears to be that the density function $g_n(\hat{\theta}; \theta, k)$, written more simply as $g_n(\hat{\theta}; \theta)$, for the ML estimates induced by $f(\mathbf{x}; \theta, k)$ should assign the same density to all ML estimates within the interval, which also is to include a point θ. We discuss the problem first for one-dimensional parameters, $k = 1$.

Let $B(\hat{\theta})$ be an interval centered at $\hat{\theta} = \hat{\theta}(\mathbf{x})$, whose length is denoted by $|B(\hat{\theta})|$. Consider the problem:

$$\max_{B(\hat{\theta})} |B(\hat{\theta})| \min_{\theta \in B(\hat{\theta})} g_n(\theta; \hat{\theta}). \qquad (6.1)$$

Write the solution as $\hat{B}(\hat{\theta})$. For simplicity we assume $g_n(\theta; \hat{\theta})$ to be unimodal for all $\hat{\theta}$, which it will be for n greater than some number if the CLT holds and causes $g_n(\theta; \hat{\theta})$ to converge in law to a gaussian density function. Then the minimizing θ falls at both ends of $B(\hat{\theta})$, where the values of $g_n(\theta; \hat{\theta})$ are equal.

Define an optimal interval estimator by

$$\mathbf{x} \mapsto \hat{B}(\hat{\theta}(\mathbf{x})), \qquad (6.2)$$

and $\hat{\theta}(\mathbf{y}) \equiv \hat{\theta}(\mathbf{x})$ if $\hat{\theta}(\mathbf{y}) \in \hat{B}(\hat{\theta}(\mathbf{x}))$: in words, $\hat{B}(\hat{\theta})$ is the interval that includes $\hat{\theta}$ and supports the rectangle within the density function $g_n(\theta; \hat{\theta})$ of maximum area $Q(\hat{\theta})$. Its probability induced by $g_n(\theta; \hat{\theta})$ is

$$\hat{P}(\hat{\theta}) = \int_{\hat{B}(\hat{\theta})} g_n(\theta; \hat{\theta}) d\theta. \tag{6.3}$$

A simplification takes place if we take $g_n(\theta; \hat{\theta})$ as a gaussian

$$g_n(\theta; \hat{\theta}, \sigma^2) = \frac{\sqrt{n}}{\sqrt{2\pi}\sigma} e^{-n(\theta-\hat{\theta})^2/(2\sigma^2)}, \tag{6.4}$$

where $\sigma = \sigma(\hat{\theta})$ may depend on $\hat{\theta}$. Then the maximum of the gaussian is at the center of the interval $\hat{\theta}$, and the length of the maximizing interval is $2\sigma(\hat{\theta})\sqrt{1/n}$. To see this let $B_{d/n}(\hat{\theta})$ denote an interval of length $2\sigma\sqrt{d/n}$ centered at $\hat{\theta}$, where d is a size parameter.

The problem (6.1) which we need to solve is

$$\max_d 2\sigma\sqrt{d/n} \times \frac{\sqrt{n}}{\sigma\sqrt{2\pi}} e^{-d/2} \tag{6.5}$$

or

$$\max_d \frac{2\sqrt{d}}{\sqrt{2\pi}} e^{-d/2}. \tag{6.6}$$

Maximizing the logarithm we get $\hat{d} = 1$, which gives the optimal interval length $2\sigma(\hat{\theta})/\sqrt{n}$, and the area of the maximum rectangle within the density function is given by

$$Q = \sqrt{2/\pi} e^{-1/2}, \tag{6.7}$$

which is seen not to depend on $\hat{\theta}$.

In addition we also calculate the probability

$$\hat{P}(\hat{\theta}) = \hat{P}_1(\hat{\theta}) = \int_{B_{1/n}(\hat{\theta})} g_n(\theta; \hat{\theta}) d\theta = \mathrm{erf}(1/\sqrt{2}) \approx 0.68, \tag{6.8}$$

where $\mathrm{erf}(x)$ denotes the error function.

An important k-dimensional parameter case is $k = 2$, in which the interval is generalized to a quadrangle, where the sides are straight but the angles are not necessarily 90 degrees. The area of the quadrangle is the sum of that of two triangles that partition it. Let B be a quadrangle that includes the point $\hat{\theta} = \hat{\theta}(\mathbf{x})$. The interval estimator is now defined as the solution to the problem: find the quadrangle B which supports the parallelogram within $g_n(\theta; \hat{\theta})$ of maximum volume

$$Q(\hat{\theta}) = \max_{B:\hat{\theta}\in B} |B| \min_{\theta\in B} g_n(\theta;\hat{\theta}), \tag{6.9}$$

where $|B|$ denotes the area of a quadrangle. Again, for unimodal density functions $g_n(\theta;\hat{\theta})$ the four sides of the parallelogram orthogonal to the supporting quadrangle are of equal length so that the upper ends touch the density function.

The generalization to any k is straightforward, except that the volume of the maximizing hyperquadrangle will be harder to calculate, and we do not do it except in the most important case where $g_n(\theta;\hat{\theta})$ is gaussian,

$$g_n(\theta;\hat{\theta}) = \frac{n^{k/2}|J_n(\hat{\theta})|^{1/2}}{(2\pi)^{k/2}} e^{-\delta' n J_n(\hat{\theta})\delta/2} \tag{6.10}$$

$$= \frac{n}{(2\pi)^{k/2}} \prod_1^k \lambda_j e^{-n/2 \sum_-^k \lambda_j \delta_j^2}, \tag{6.11}$$

where $\delta = \theta - \hat{\theta}$ and λ_j are the eigenvalues of the Fisher information matrix, (4.58),

$$J_n(\theta) = \frac{1}{n} E_\theta \left\{ \frac{\partial^2 \ln 1/f(X^n;\theta,k)}{\partial \theta_j \partial \theta_k} \right\}. \tag{6.12}$$

We assume that $J_n(\theta) \to J(\theta)$ for all θ. We indicate the transpose by a prime.

If the CLT holds for the models in the class $\mathcal{M}_k$, then with the change of variables $\delta = \sqrt{n} J_n(\hat{\theta})^{1/2}(\theta - \hat{\theta})$,

$$g_n(\delta;0) \to \frac{1}{(2\pi)^{k/2}} e^{-\delta'\delta/2}. \tag{6.13}$$

Then approximating $g_n(\theta;\hat{\theta})$ by the gaussian, (6.10), the maximizing generalized interval is a k-dimensional hyperrectangle of maximum volume centered at $\hat{\theta} = \hat{\theta}(\mathbf{x})$, which supports the maximum volume parallelogram whose sides are of equal length and touch the gaussian density function. In fact, consider a hyperellipsoid $D_{d/n}(\hat{\theta})$ in the bounded parameter space

$$(\theta - \hat{\theta})' J_n(\hat{\theta})(\theta - \hat{\theta}) = d/n, \tag{6.14}$$

centered at point $\hat{\theta}$, where d is a parameter determining the volume of the hyperellipsoid. Let $B_{d/n}(\hat{\theta})$ be the largest hyperrectangle within $D_{d/n}(\hat{\theta})$. Its orientation is determined by the eigenvectors of $J_n(\hat{\theta})$, see Figure 6.1.

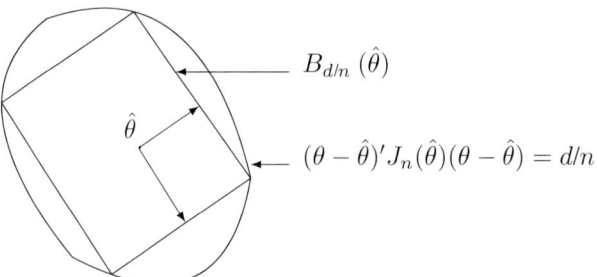

Figure 6.1 A hyperrectangle as hyperinterval.

Its volume is

$$|B_{d/n}(\hat{\theta})| = \left(\frac{4d}{k}\right)^{k/2} |J_n(\hat{\theta})|^{-1/2} = 2^k \prod_{j=1}^{k} \mu_j, \qquad (6.15)$$

where $\mu_j = \sqrt{d/(k\lambda_j)}$ is one half of the jth side length of the hyperrectangle, and λ_j is the corresponding eigenvalue of $J_n(\hat{\theta})$. We get this formula by maximization of the volume $2^k \prod_{i=1}^{k} a_i$ of hyperrectangles with side length $2a_i$ and corners on the surface of the hyperellipsoid.

The sides of the parallelogram supported by the hyperrectangle $B_{d/n}(\hat{\theta})$ are given by (6.10) as

$$\frac{n^{k/2}|J_n(\hat{\theta})|^{1/2}}{(2\pi)^{k/2}} e^{-d/2}, \qquad (6.16)$$

and its volume is given by this side length times the volume of the hyperrectangle. The maximization problem, generalizing (6.9), amounts to

$$\max_d \left(\frac{2d}{\pi k}\right)^{k/2} e^{-d/2}. \qquad (6.17)$$

The solution is $\hat{d} = k$, which gives the volume of the optimal hyperrectangle as $(4/n)^{k/2}|J_n(\hat{\theta})|^{-1/2}$ and the constant volume of the maximum parallelogram $Q = (2/(e\pi))^{k/2}$, supported by the hyperrectangle within the density function. Further, the probability of the hyperrectangle is

$$\hat{P}_k(\hat{\theta}(\mathbf{x})) = \int_{B_{k/n}(\hat{\theta})} g_n(\theta; \hat{\theta}, k) d\theta = (\text{erf}(1/\sqrt{2}))^k, \qquad (6.18)$$

where $\text{erf}(x)$ denotes the error function, and $\text{erf}(1/\sqrt{2}) \approx 0.68$.

6.2 Maximum capacity partition

In addition to the optimal interval estimators, which give the interval that includes the ML estimate, we want to study estimators which could be regarded as maximum capacity estimators in the same sense as the similar point-wise estimators. This requires us to define the capacity and the estimator that maximizes it, which in turn requires a partition of the parameter space into intervals. We study in detail the one-dimensional parameter case. As it happens there are three partitions and the associated capacities, which define estimators that maximize the capacity. The first is the trivial one, defined by the partition with a single equivalence class, the entire parameter space, with the maximum capacity 0. The second is the zero-interval partition studied earlier whose equivalence classes are defined by the ML estimates, with the maximum capacity $\log \hat{C}_k$. The third is relevant for the proper non-zero intervals to be defined by an algorithm. For simplicity we describe the algorithm for the one-dimensional parameter space $\Omega = [a, b]$, and extend it to two-dimensional parameters in the last subsection of this chapter. As stated in the preceding section it can be generalized to any dimensional parameters, except for computational problems.

The algorithm partitions the parameter space $\Omega = [a, b]$ into intervals $(c_i, c_{i+1}]$, except the left-most, which is $[c_i, c_{i+1}]$, and we denote by $B(\theta_i)$ an interval centered at the point θ_i. The algorithm is easier to understand if the density functions $g_n(\hat{\theta}; \theta)$ are unimodal for all θ. This is not hard to satisfy for by the CLT the density function $g_n(\hat{\theta}; \theta)$, for each θ can be sufficiently well approximated by a gaussian to be unimodal. The algorithm will find a set of points $\theta_0, \theta_1, \ldots, \theta_m$ defining intervals.

Algorithm 6.1

1. Start with a point θ_0; we discuss its selection later. Find the interval $B(\theta_0) = (c_0, c_1]$ that includes θ_0 and supports the rectangle within $g_n(\hat{\theta}; \theta_0)$ of maximum area:

$$Q(\theta_0) = \max_{c_0, c_1} (c_1 - c_0) g_n(c_0; \theta_0). \qquad (6.19)$$

Notice that $g_n(c_0;\theta_0) = g_n(c_1;\theta_0)$ for unimodal density functions $g_n(\hat{\theta};\theta)$, and otherwise solve

$$Q(\theta_0) = \max_{c_0,c_1}(c_1 - c_0) \min_{c:c_0 \leq c \leq c_1} g_n(c;\theta_0). \tag{6.20}$$

2. Find $\theta = \theta_1$ and $c = c_2$ as solutions to

$$\max_{\theta,c}(c - c_1) g_n(c;\theta) = Q(\theta_1) \tag{6.21}$$

in the case of a unimodal density function, and put $B(\theta_1) = (c_1, c_2]$. If not unimodal, solve

$$\max_{\theta,c}(c - c_1) \min_{\alpha:c_1 \leq \alpha \leq c} g_n(\alpha;\theta) = Q(\theta_1). \tag{6.22}$$

3. Continue finding $\theta_2, \theta_3, \ldots$ until $c_j = b$.
4. Continue by finding $\theta_{-1}, \theta_{-2}, \ldots$ and decreasing boundaries $c_{-1}, c_{-2}, \ldots$ until $c_{-k} = a$ by solving

$$\max_{\theta,c}(c_0 - c) \min_{\alpha:c \leq \alpha \leq c_0} g_n(\alpha;\theta) = Q(\theta_{-1}). \tag{6.23}$$

Put $B(\theta_{-1}) = (c_{-1}, c_0]$, $B(\theta_{-2}) = (c_{-2}, c_1], \ldots$.

The algorithm defines the partition $\hat{\Lambda}_n = \{B(\theta_i)\}$.

Rewrite

$$Q(\theta_i) = \max_{B(\theta_i)} |B(\theta_i)| \min_{\theta \in B(\theta_i)} g_n(\theta;\theta_i), \tag{6.24}$$

where the minimizing θ is at both ends of the interval. The algorithm defines a probability distribution for the intervals by,

$$\hat{Q}(\theta_i) = Q(\theta_i)/\hat{C}[n,\theta_0], \tag{6.25}$$

$$\hat{C}[n,\theta_0] = \sum_j Q(\theta_j), \tag{6.26}$$

where $\log \hat{C}[n,\theta_0]$ is the capacity.

Unlike in the case with point-wise ML estimators, the capacity for interval estimation depends on the initial point θ_0. If $g_n(\theta;\theta_i)$ is a uniformly continuous function of θ at all θ_i, the dependency is reduced with increasing n. At any rate the logically best choice is the one that maximizes the capacity: $\max_{\theta_0} \log \hat{C}[n,\theta_0] = \log \hat{C}[n]$ under the equivalence $\theta \equiv \theta'$ if both belong to the same equivalence class.

By Lemma 4.1 the given partition satisfies the same necessary condition for the maximum as follows. Any alteration of the equivalence classes in $\hat\Lambda_n$ small enough not to change their number reduces $\log \hat C[n]$, calculated at the optimum θ_0, and hence $\hat Q(\theta_i)$. It also gives the maximum amount of information an interval estimator with the number of the equivalence classes $\hat m_n$ obtains about the density functions $\{g_n(\hat\theta;\theta) : \theta \in \Omega\}$.

To summarize this section we give the maximum capacity interval estimator:

$$\mathbf{x} \mapsto B(\theta_i(\mathbf{x})), \qquad (6.27)$$

where $B(\theta_i(\mathbf{x}))$ is the equivalence class in $\hat\Lambda_n$, defined by Algorithm 6.1, that includes $\hat\theta(\mathbf{x})$.

6.3 Error probability

For a finite family of models $\mathcal{P} = \{p(\mathbf{x};\theta_i) : i = 1, 2, \ldots, m\}$ define the *mean separation* $\bar P$ as

$$\bar P = \frac{\hat C}{m}, \qquad (6.28)$$

$$\hat C = \sum_i P(\theta_i), \qquad (6.29)$$

$$P(\theta_i) = \int_{A_i} p(\mathbf{x};\theta_i) d\mathbf{x}, \qquad (6.30)$$

$$A_i = \{\mathbf{x} : p(\mathbf{x};\theta_i) \geq p(\mathbf{x};\theta_j), j \neq i\}: \qquad (6.31)$$

in words, the classes A_i partition the space X of the data sequences by the ML estimates of the models as shown in Figure 3.1. If we imagine that one of the m models generates the data $\mathbf{x}$, and we estimate it by the parameter θ_i, then the probability of the estimation error is $1 - P(\theta_i)$, whose average, given the family, is seen to be minimized by $\bar P$:

$$1 - \bar P = \frac{1}{m}\sum_i (1 - P(\theta_i)). \qquad (6.32)$$

Let $\Lambda(m) = \{B(\theta_1), \ldots, B(\theta_m)\}$, for $B(\theta_i) = \{\hat\theta(\mathbf{x}) : \mathbf{x} \in A_i\}$, denote a partition of the parameter space Ω, induced by the family or the set of the m points $\{\theta_i\}$. We are interested in partitions, of which there can be an immense number, for which the $P(\theta_i)$ are large, because then the probabilities of estimation errors are small.

A desirable property of a partition $\Lambda(m)$ for each m is to minimize the worst case estimation error

$$\min_{\Lambda(m)} \max_i (1 - P(\theta_i)). \tag{6.33}$$

However, more is true of such minimizing partitions $\bar{\Lambda}(m)$: for a given set of density functions $g_n(\hat{\theta}; \theta)$, which are continuous at all θ in a compact set Ω, the minimizing partition is such that $P(\theta_i)$ is constant P_m for all i, some P_m that depends only on m. Indeed, if that is not the case, the minimum worst case error probabilities, say $P_E = 1 - P(\theta_k)$ for $k = k_1, k_2, \ldots$ where they are equal, could be decreased by a small increase of their intervals and a reduction of the other intervals where the error probabilities are larger than the minimum. This would contradict the supposed minimizing partition. This is an example of the estimation problem where the ML estimates also maximize the mutual information as discussed in Part I.

The partition $\hat{\Lambda}_n$ has a second optimization property for the gaussian family $g_n(\hat{\theta}; \theta)$, (6.4), namely that it is the only one for which both $\hat{Q} = \sqrt{2/\pi}e^{-1/2}$ and $\hat{P} = \text{erf}(1/\sqrt{2})$ minimize the worst case error probability of the distributions $\hat{Q}(\theta_i)$ and $\hat{P}(\theta_i)$, respectively. In view of the CLT this holds approximately even when $g_n(\hat{\theta}; \theta)$ is not gaussian. Except possibly for the intervals that include the two endpoints a and b, both of these distributions are about the same, namely, $1/\hat{m}_n$. This, together with the uniqueness of the partition $\hat{\Lambda}_n$ of size $\hat{m}_n$, gives the family $\{g_n(\hat{\theta}; \theta_i) : i = 1, 2, \ldots, \hat{m}_n\}$ for each n a powerful justification, which we call *optimally distinguishable*. The results generalize easily to the models in $\mathcal{M}_k$ for any k and so does the notion of optimal distinguishability. The only difficulty is the calculation of the size $\hat{m}_n$, except for large n, which we discuss next for the case $k = 1$.

6.3.1 Asymptotic distinguishability

We are interested in the partition $\hat{\Lambda}_n$ for large n and for the normally distributed family (6.4). Writing $g_n(\theta) = g_n(\theta; \theta, \sigma^2(\theta))$ for simplicity, let for d

$$\hat{G}_d(\theta_i) = \int_{B_{d/n}(\theta_i)} g_n(\theta, \sigma^2) d\theta, \tag{6.34}$$

which gives $\sum_1^{m_n(d)} \hat{G}_d(\theta_i) = \hat{C}_n$. Further, consider the upper bound for rectangles of constant area $Q_d = \sqrt{2d/\pi}e^{-d/2}$, (6.7),

$$g_n(\theta_i)|B_{d/n}(\theta_i)| = Q_d e^{d/2} = \sqrt{2d/\pi}, \qquad (6.35)$$

whose top goes through the point $g_n(\theta_i)$. We see at once that for large n its volume is very close to $\hat{G}_d(\theta_i)$, and we get an approximation of $m_n(d)$. In fact, if $1/\sigma(\theta)$ is differentiable at $\theta \in B_{d/n}(\theta_i)$, the expansion in a Taylor series gives

$$g_n(\hat{\theta}) = g_n(\theta_i) + g'_n(\theta_i)\delta + O(\delta^2), \qquad (6.36)$$

where $\delta = \hat{\theta} - \theta_i$ and $g'_n(\theta_i)$ is the first derivative at the point θ_i. From (6.34) then

$$\hat{G}(\theta_i) = \sqrt{2d/\pi} + g'_n(\theta_i)\int_{-\sqrt{d/n}}^{\sqrt{d/n}} \delta\, d\delta + O((d/n)^{3/2}2) \qquad (6.37)$$

$$= \sqrt{2d/\pi} + O((d/n)^{3/2}). \qquad (6.38)$$

Further, $\hat{C}_n = m_n(d)\sqrt{2d/\pi}\,[1 + O((d/n)^{3/2})]$ and

$$\log m_n(d) = \log \hat{C}_n + \log\sqrt{\pi/(2d)} + O((d/n)^{3/2}). \qquad (6.39)$$

Consider the ideal code length $\log 1/f(x^n;\theta_i,1)$, which using a Taylor series expansion gives

$$\log 1/f(x^n;\theta_i,1) = \log 1/f(x^n;\hat{\theta}(x^n),1) + \frac{1}{2}n\delta^2/\sigma^2(\theta_i) + O((d/n)^{3/2}). \qquad (6.40)$$

Then by adding the code length for θ_i, which is $\log m_n(d)$, we get the total code length for x^n:

$$\log 1/f_d(x^n;\theta_i,1) + \log m_n(d) = \log 1/\hat{f}(x^n;1) + \frac{1}{2}n\delta^2/\sigma^2(\theta_i)$$
$$+ \log\sqrt{\pi/(2d)} + O((d/n)^{3/2}). \qquad (6.41)$$

The worst case δ is on the boundary of the interval $B_{d/n}(\theta_i)$, or $\delta = \sigma(\theta_i)\sqrt{d/n}$. Minimization of the ideal code length at the worst case δ gives $\hat{d} = 1$. This means that the so obtained partition differs from $\hat{\Lambda}_n$ only by the choice of the initial point θ_0, the effect of which we ignore. Further, the worst case excess due to the quantization over the non-quantized code length is given to a good approximation by $0.5 + \log\sqrt{\pi/2} = 0.825$, or less than one bit for the entire string.

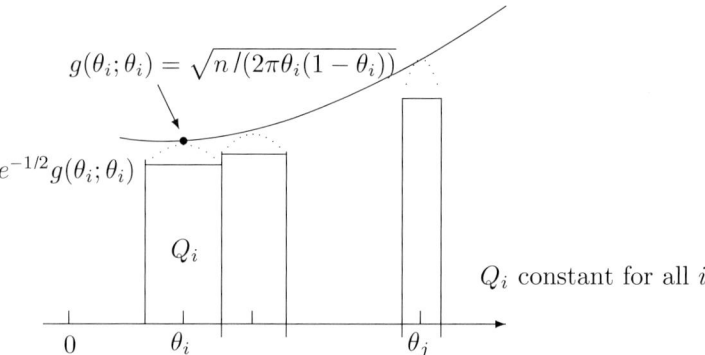

Figure 6.2 Bernoulli models.

The conclusion is that the family of optimally distinguishable models $\{f(x^n;\theta_i;1) : i \leq m_n(1)\}$ loses hardly anything in the code length to the full uncountable family.

Further, the mean separation for the family $\{g_n(\hat{\theta};\theta_i)\}$ for $\theta_i \in \hat{\Lambda}_n$, consisting of normally distributed density functions, is constant, as given by (6.8). Hence the maximum capacity $\log \bar{C}_n$ for intervals $B(\theta_i)$, estimated by the equivalence $\hat{\theta}(\mathbf{x}) \equiv \theta_i$, is given by (6.8) as

$$\bar{C}_n = \hat{m}_n \bar{P} = \hat{C}_n [\sqrt{\pi/2} + O((1/n)^{3/2})] \operatorname{erf}(1/\sqrt{2}) \approx 0.85 \hat{C}_n, \qquad (6.42)$$

which is less than $\hat{C}_n$ as it should be.

As an illustration of the optimally distinguishable models we consider the Bernoulli class in Figure 6.2.

The optimally distinguishable models for $\mathcal{M}_k$ were found by asymptotic means in [45]. The compact parameter space was partitioned into hyperrectangles, defined by the empirical Fisher information matrix. The difficulty is that such rectangles cannot partition the parameter space unless the Fisher information matrix is constant. This led into an awkward approximation of it by piece-wise constant matrices. The optimal value $\hat{d} = k$ was found by minimization of the worst case code length within the equivalence classes. It is clear that Algorithm 6.1 generalizes to $k > 1$ for any n, except for the details, as we show for $k = 2$ in Section 6.3.2. Hence, the set of (non-asymptotic) optimally distinguishable models can be extended to any k. We now have a different non-asymptotic interpretation and derivation based on the main postulate: maximize the density value a model assigns to the observed data under the restriction that the estimated parameters that fall within the optimal hyperintervals are equivalent. The second essentially equivalent interpretation is to maximize the estimation information of the family $g_n(\hat{\theta}; \theta_i)$ under the same equivalence restriction.

The notion of distinguishability was inspired by Balasubramanian [1], who showed that the "regret," which asymptotically amounts to the maximum capacity $\log \hat{C}_k$, can be viewed as the number of *distinguishable* models. The main objective for him was to obtain a formal way to define models that can be distinguished from those that are too close to each other. He managed to do this without partitions. Instead he used an asymptotic form of Stein's lemma, which is not applicable to finite families.

6.3.2 Partition algorithm for $k = 2$

In this last subsection we generalize Algorithm 6.1 to $k = 2$, which has applications to image processing. For simplicity we consider the case where the density functions $g_n(\hat{\theta}; k)$ are approximated by normal distributions (6.10) of mean θ and covariance matrix $\Sigma(\theta) = J_n^{-1}(\theta)$, given by the inverse of the Fisher information matrix. In general, the covariance matrix varies as a function of θ in the parameter space Ω, which we take to be compact. The equivalence classes $B(\theta_i)$ will be quadrangles, where the angles are not orthogonal, except in the very first.

Algorithm 6.2

1. Start with a point θ_0, which can be any point in Ω. Find the rectangle $B(\theta_0)$ centered at θ_0, which supports the parallelogram within $g_n(\hat{\theta}; \theta_0, \Sigma(\theta_0))$ of maximum volume:

$$Q(\theta_0) = \max_{B(\theta_0)} |B(\theta_0)| \min_{\theta \in B(\theta_0)} g_n(\theta; \theta_0, \Sigma(\theta_0)), \qquad (6.43)$$

where $|B|$ denotes the area of a quadrangle, here a rectangle.

2. Find θ_1 and quadrangle $B(\theta_1)$, which shares a common edge with $B(\theta_0)$ and supports the parallelogram of maximum volume within $g_n(\theta; \theta_1, \Sigma(\theta_1))$

$$\max_{B(\theta_1)} |B(\theta_1)| \min_{\theta \in B(\theta_1)} g_n(\theta; \theta_1, \Sigma(\theta_1)) = Q(\theta_1). \qquad (6.44)$$

3. Continue finding $\{\theta_i\}$ and $\{B(\theta_i)\}$ for $i = 2, 3, \ldots$ along all directions where any two adjacent quadrangles share a common edge until the boundary; in Figure 6.3 the four arrows indicate the four directions of these quadrangles.

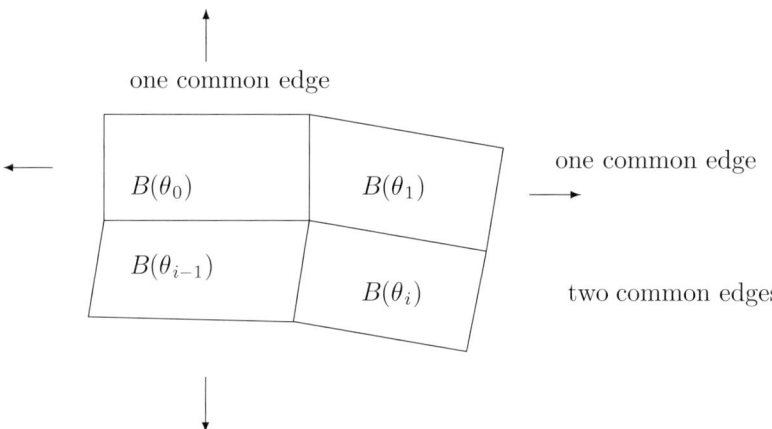

Figure 6.3 Two-dimensional partition by Algorithm 6.2.

4. Next, find the remaining centers θ_j and quadrangles $B(\theta_j)$ which share two common edges with the preceding quadrangle and support the parallelograms of maximum volume within $g_n(\theta; \theta_j, \Sigma(\theta_j))$.

The area of a quadrangle is the sum of the areas of two triangles which partition the quadrangle. Also the minimizing $\theta \in B(\theta_i)$ is in all corners of the maximizing quadrangle, and all the four edges of the maximum volume parallelogram that touch the density function $g_n(\theta; \theta_i, \Sigma(\theta_i))$ have the same length. Notice, however, that the lengths of the edges of the adjacent parallelograms need not be equal. Finally, the partition depends on the center of the initial rectangle, which could be taken as an extremum of the peaks $g_n(\theta; \theta, \Sigma(\theta))$. This partition solves a difficulty in attempts to construct a partition into rectangles, which forces the edges to be curvilinear.

The algorithm defines the probability distribution for the intervals by

$$\hat{Q}(\theta_i) = \frac{Q(\theta_i)}{\sum_j Q(\theta_j)}. \tag{6.45}$$

It also defines a histogram estimation of the density function $g_n(\theta; \theta, \Sigma(\theta))$:

$$\hat{h}(\hat{\theta}) = \hat{Q}(\theta_i)/|B(\theta_i)|, \tag{6.46}$$

where $\hat{\theta} \equiv \theta_i$.

7 Hypothesis testing

Customarily, hypothesis testing is about the problem of finding out if some effect of interest, like the curing power of a drug, is supported by observed data, or if it is a random "fluke" rather than a real effect of the drug. To have a formal way to test this a certain probability hypothesis is set up to represent the effect. This can be done by selecting a function of the data, $s(x^n)$, called a *test statistic*, the distribution of which can be calculated under the hypothesis. Frequently parametric probability distributions have a special parameter value to represent randomness, like $1/2$ in Bernoulli models for binary data and 0 for the mean of normal distributions, and the no effect case can be represented by the single model such as $f(s(x^n); \theta_0)$, called the *null hypothesis*. The test statistic is or should be in effect an estimator of the parameter, and if the data cause the test statistic to fall in the tail, into a so-called *critical region* of small probability, say 0.05, the null hypothesis is rejected, indicating by the double negation that the real effect is not ruled out. Fisher called the amount of evidence for the rejection of the null hypothesis the *statistical significance* on the selected level 0.05, which is frequently misused to mean that there is 95 percent statistical significance to accept the real effect. As a matter of fact, there is nothing in current hypothesis testing to justify any degree of statistical evidence to accept any hypothesis. Neither, for that matter, is the level alone enough to measure statistical significance. Consider the uniform distribution. Any subset whose size is $5/100$ of the range could be taken as the critical region on the same level, and since there is no way to prefer one over the others the level as a measure of statistical significance for the rejection is meaningless in general.

7.1 General plan

The aim in this section is to investigate in general terms how hypothesis testing can be founded on optimal estimation and the associated ideas, and to outline how a more symmetric testing without the lopsided null hypothesis can be achieved. More than in estimation this requires changing and modifying the customary approach, above all sharpening the requirements for the test statistics and their distributions.

The basic assumption is that a hypothesis is either a model in a class such as $\mathcal{M}_k$, or it has to be estimated such that it falls in a subset of the class of models when we are interested in some broader property of the models. When it is a specific model, defined by a single parameter, the hypothesis is called *simple*, while when it is to be estimated it is customarily called *composite*, instead of the more logical name *complex*. Since we are also interested in the statistical evidence to accept a hypothesis, whether it expresses a negative statement or not, we drop the superfluous and confusing prefix "null." Instead, we regard testing as the problem of finding out if a given data string belongs to the set of *typical* strings of the model or models defining the hypothesis, i.e. the most frequently generated strings, if we were to repeat the sampling, or if it is a rare exceptional atypical string, which in the terminology inspired by Appendix A could be called *random* relative to the hypothesis. The meaning is that the string lacks the statistical properties expressed by the hypothesis.

The basic way to define typicality is in terms of the code length in that it can be compressed to the length of the typical strings of the hypothesis. Although the code length as the criterion is not sharp enough for hypothesis testing, in which we only need to "estimate" typicality rather than the exact model generating the string, it will give a general idea of what is needed. We could call a string $\mathbf{x}$ typical of a model $f(\mathbf{y}; \theta)$ in $\mathcal{M}_k$, if

$$\log 1/f(\mathbf{x}; \hat{\theta}(\mathbf{x})) \leq \log 1/f(\mathbf{x}; \theta) < \log 1/f(\mathbf{x}; \hat{\theta}(\mathbf{x})) + \log \hat{C}_k, \quad (7.1)$$

where the first inequality is automatically satisfied but included to show the amount by which the code length of the model exceeds the unavailable ideal code length, while the second tells how much shorter it is than the optimally estimated "universal" code length, which is available for any string.

For rejection of a hypothesis, a string could be called *random* relative to the model if

$$\log 1/f(\mathbf{x};\theta) > \log 1/f(\mathbf{x};\hat{\theta}(\mathbf{x})) + \log \hat{C}_k. \qquad (7.2)$$

Finally, a string is random relative to the entire model class $\mathcal{M}_k$, if $\min_{\theta \in \Omega^k} \log 1/f(\mathbf{x};\theta)$ is not shorter than the code length of the string $\mathbf{x}$ when encoded just as a binary string without use of any model in the class.

The code length, however, is not adequate for hypothesis testing, because its distribution is generally very flat, and we need a test statistic with a "peaked" distribution which allows a sharp separation of the typical strings from the atypical ones. There are also other reasons, as will be discussed below. Since the test statistic is used as a broad characterization of the data strings, its distribution should retain as much information about typicality as possible. We spell out below the requirements of a good test statistic, which seem to have been overlooked in traditional testing, and select two which cover a fairly good part of testing cases.

The interpretation of the hypotheses as models without any reference to "truth" has the important consequence that we can test for acceptance and avoid the logical difficulty arising from the fact that we make the right decision if we reject the hypothesis as the generator of the string regardless of the data. This is the case even if instead of the true model we settle for the best model in the set of all computable distributions, because it cannot be found in a computable manner by any algorithm; see the Appendix A. When the hypotheses are just models both the rejection and the acceptance can be done if we settle for tests of typicality, which fortunately can be done. To emphasize the difference, suppose we consider binary data like 01010101010101..., and the hypothesis is the Bernoulli model with the probability of each symbol being 1/2. By the tests described below the hypothesis would be accepted. We usually regard typical strings from the Bernoulli model to be more "random" looking than this string, but the fact is that this string is just as likely to occur as any other string in which the numbers of 0s and 1s are equal or nearly so.

In this theory our basic means of fitting models to data is based on maximization of the probability (or density) a model assigns to data, or, equivalently, minimization of the code length. Therefore, since we are discussing the testing of models with a fixed number of parameters, the optimal ML estimates could be considered to provide the primary test statistic. However, the task is more complex, for we think of data as being generated not by any distribution but by some physical machinery, which we model by a distribution. Consequently we regard data to consist of at least three parts: First there is the relevant useful part, which we try to capture by the models. Then there are the "local" random variations, ordinarily called "noise," and generated by something we do not know, which is the very reason we select probabilistic models rather than deterministic models of a function type. The third part also consists of random variations, but they are of slowly varying drift type, which are harder to detect, and whose origin we do not understand. The problem is to quantify and measure these to have a basis for testing. The optimal interval estimation goes a long way in dealing with the fast variations, the noise, in effect by providing the optimal precision for the real-valued parameters. This was achieved in terms of the Fisher information matrix by taking into account the sensitivity of the code length with respect to the variations of the parameters.

For the long term variations we need another estimate and technique to decide whether the data are or are not typical for the hypothesis, which is either simple, i.e. represented by a single model, or composite. If the hypothesis is simple the task is to test whether the shape of the density function, defined by the model, agrees with the data. This can be done in terms of the cumulative distribution, which calls for the data to be represented by histograms, which involves a loss of information though the information lost is mostly about the short term variations. In the case of composite hypotheses there is an added difficulty. The best model is defined by the ML estimate, which minimizes the code length of each data string, typical or not. After all, there are lots of strings all giving the same ML estimate, some typical for the estimated model others atypical, and the critical decision must be made on the second type of test statistic.

Although no sharp boundary between typicality and atypicality exists a good test statistic does make it meaningful to consider the three-way decision:

1. rejection of the hypothesis if the data are to be considered random relative to it, or a "fluke," not having the statistical property represented by the hypothesis;

2. acceptance if the data are judged to be typical of the posited hypothesis; and, finally,

3. neither if there is not enough evidence to decide the matter.

The last step is not new, for it has been used in sequential testing [55] for the purpose of obtaining more data when the judgment (to reject) is that a decision cannot be made. Setting fixed boundaries must be regarded as arbitrary and important only if other issues like costs and money are involved, and they add no new information or reliance to the statistical findings. Without knowing such issues we are not particularly interested in any specific decision boundaries except for acceptance, where noise can be optimally detected and its effect avoided.

7.2 Test statistics

To sharpen the typicality and relative randomness statements we need the crucial test statistic or statistics, just as in traditional testing, for which we also need the distribution to be able to make probability statements of the degree of typicality. As is the case with estimators, not all test statistics are equally good. To illustrate this important point suppose the parameter in a simple hypothesis $f(\mathbf{x}; \theta)$, determined by a fixed value, specifies the mean, and we want to test if the data conflict with it. Any observation point x_i could be used for this, for if it falls far from the mean of the hypothesis, it should cause rejection. But clearly, the arithmetic mean estimator is a better test statistic, for its distribution is more peaked. For the same reason the data themselves do not provide the best test statistics, because the hypothesis model is often flat, the more so the more data we have.

Unlike in traditional testing, in which the main and often the only requirement of a test statistic is that its distribution can be calculated, at least approximately, here we spell out the requirements for good test statistics, after which we select two test statistics. The first is the ML estimator and the second is defined by the KL distance, and together these apply in great generality to the problems of hypothesis testing.

1. A test statistic as a function of the observed data plays the role of an estimator function for parameters. It should reflect the hypothesis and incorporate the relevant statistical information in the data.
2. Its distribution, induced by the hypothesis, should assign a large probability to a small region of the typical data of the hypothesis and a small probability to the atypical data in order to separate well typicality from atypicality.
3. These features should improve with growing quantities of data.

The second important and in fact crucial requirement providing an easy separation allows us to define the region of acceptability almost as sharply as that of the rejection. The third requirement seems justified just as in estimation, because test results should be better with large quantities of data.

As an illustration of a good test statistic consider the simple hypothesis $f(\mathbf{x}; \theta, k)$ as the model of the data $\mathbf{x}$. Take the ML estimator $\hat{\theta}(\mathbf{x})$ as the test statistic, which has the density function $g(\hat{\theta}; \theta, k)$. The distribution $f(\mathbf{x}; \theta, k)$ determines $g(\hat{\theta}; \theta, k)$, which together with the test statistic retains most if not all the relevant information needed to "estimate" the degrees of typicality and atypicality of the data for the hypothesis. This is seen as follows. Let A denote an acceptance region as the set of typical data for the hypothesis. Its probability is given by

$$P(A) = \int_{\hat{\theta} = \hat{\theta}(\mathbf{x}) : \mathbf{x} \in A} g(\hat{\theta}; \theta, k) d\hat{\theta}. \tag{7.3}$$

Moreover, in our theory the estimator $\hat{\theta}(\cdot)$ satisfies the necessary condition for optimality, Theorem 4.1. It will therefore satisfy the necessary condition for $P(A)$ to be maximum for all A. If for the moment we take the complement of A as the set of atypical strings for the hypothesis, the test statistic $\hat{\theta}(\cdot)$ minimizes its probability, and we may regard it as the optimal test statistic. It has the further advantage that if it satisfies the CLT its density function is approximately gaussian, which confirms the crucial second requirement of a good test statistic.

7.3 Characteristic histograms

If we apply the partition Algorithm 6.1 to parametrically defined density functions $f(y; \theta, k)$ we get what could be called a *characteristic* histogram approximation of them. Start with a model $f(y;\theta,k)$, extended to sequences $\mathbf{y}$ either by independence $f(\mathbf{y};\theta,k) = \prod_t f(y_t;\theta,k)$, or in the case of conditional models $f(y_{t+1}|y^t;\theta,k)$, by the product $f(\mathbf{y};\theta,k) = \prod_t f(y_{t+1}|y^t;\theta,k)$. Then calculate the ML estimate $\hat{\theta}(\mathbf{x})$ from the available data $\mathbf{x} = x^n$, which defines the estimated model for the symbols $f(y;\hat{\theta}(x^n),k)$ in the case of independence, or $f(y|x^n,\hat{\theta}(x^n),k)$, the latest conditional, otherwise.

We describe next a modification of the partition Algorithm 6.1 to get a partition of a compact proper subset of the support of a density function $f(y;\theta) = f(y;\theta,k)$, where the fixed index k is dropped to simplify the notations. We also write θ for the parameter, whether estimated $\hat{\theta}(x^n)$ or not. The partition is defined by the break points c_i, which are not only determined by the density function but in fact characterize it. We assume the density functions to be unimodal, which can be generalized to other shapes by a further modification of the algorithm. An exception is the uniform density function, where the algorithm will give just one equivalence class, which does not characterize anything. The break points for the uniform and nearly uniform density functions could be taken as done traditionally such that they define equal-length bins, whose number, however, should be optimized by the MDL principle.

Algorithm 7.1

1. Start by finding the interval $(c_0, c_1]$ that includes the largest point of $f(y;\theta)$ and supports the rectangle within $f(y,\theta)$ of maximum area:

$$A(0) = \max_{c_0,c_1}(c_1 - c_0) \min_{c:c_0 \leq c \leq c_1} f(c;\theta). \qquad (7.4)$$

2. Continue solving

$$\max_{c_2}(c_2 - c_1) \min_{c:c_1 \leq c \leq c_2} f(c;\theta) = A(1). \qquad (7.5)$$

3. Continue finding $c_3, c_4, \ldots$ until $A(m) \leq \epsilon$, where ϵ is so small that A_m is ignorable.

4. Continue in the same manner by finding decreasing boundaries $c_{-1}, c_{-2}, \ldots$ until $A(-j) \leq \epsilon$, and at the end relabel the break points from left to right by positive indices $\{c_1, c_2, \ldots, c_{m+j}\}$.

Define the distribution q over the characteristic intervals by the probabilities

$$q_i = A_i / \sum_j A_j, \qquad (7.6)$$

which in turn defines a histogram approximation of the model density function $f(y; \theta)$:

$$\hat{h}(y; \theta) = q(i)/(c_{i+1} - c_i), \qquad (7.7)$$

where $(c_i, c_{i+1}]$ is the interval where y falls.

The histogram approximation provides an idea of how well the ML model fits the data and actually represents the intended model in the family. After all, given the data sequence, the ML estimate $\hat{\theta}(\mathbf{x})$ defines some model in the family, but we have little idea of whether the shape of the density function defined by the estimated parameter fits the data well or not. To get a measure of the fit we compare the histogram approximation (7.7) with the histogram defined by the bin count ratios n_i/n, where n_i is the number of the observed data points falling in the interval $(c_i, c_{i+1}]$. The comparison is done using the KL distance as the test statistic as described in the next section, which is exact for all data and quite different from the traditional asymptotic χ^2 approximation.

Just like the interval estimators, by Algorithm 6.1 the characteristic histograms are optimal for each fixed support among all histogram estimators with the same number of bins in that they satisfy the necessary conditions for all data.

The break points also define the so-called expected bin probabilities:

$$p_i = \int_{c_i}^{c_{i+1}} f(y; \theta) dy. \qquad (7.8)$$

By Theorem 6.1 they do not differ from (7.6) by much.

The algorithm can be applied even to discrete models.

Example We give an example by applying the algorithm to the gaussian density function $f(y; 0, \sigma^2)$ of mean 0 and variance σ^2. By symmetry start with $c_0 = 0$ and find recursively the positive break points $c_0 < c_1 < c_2 < \cdots < c_m$ and the intervals $(c_i, c_{i+1}]$ that support the rectangles within $f(y; 0, \sigma^2)$ of maximum area by solving

$$\max_c \frac{c - c_i}{\sqrt{2\pi}} e^{-c^2/(2\sigma^2)}. \qquad (7.9)$$

Taking the logarithm and putting the derivative to zero gives $1/(c - c_i) = c^2/(2\sigma^2)$, with the solution

$$c_{i+1} = \frac{c_i}{2}\left(1 + \sqrt{1 + 4\sigma^2/c_i^2}\right). \tag{7.10}$$

If we take a few positive values of c_i, say three, we have $c_1 = \sigma$, $c_2 = 1.62\sigma$, $c_3 = 2.1\sigma$, and we have the additional two negative values $c_{-2} = -c_2$ and $c_{-3} = -c_3$.

For $i > 0$ we get

$$A_{i+1} = \frac{(c_{i+1} - c_i)}{\sigma\sqrt{2\pi}} e^{-(c_{i+1}^2/(2\sigma^2))}, \tag{7.11}$$

while $A_1 = 2c_1/(\sigma\sqrt{2\pi})e^{-1/2}$. Because all the break points are proportional to σ, it cancels and all A_i are independent of σ. The normalized histogram probabilities are given in (7.6).

We see that the intervals c_i shrink with increasing i and the rectangle areas shrink even more rapidly. This means that only a few intervals are needed in the characteristic histogram $\hat{h}(y)$ to give an efficient approximation of the gaussian distribution function. The five probabilities p_i in the example are approximately

$$0.04, 0.1, 0.68, 0.1, 0.04, \tag{7.12}$$

which are close to the corresponding probabilities $q_i \in \{0.03, 0.1, 0.73, 0.1, 0.03\}$. The calculation of the probabilities p_i in general parametric models may be cumbersome, and for hypothesis testing it could be avoided by replacing them by the probabilities q_i.

Example Take the Laplace distribution $f(y) = (\lambda/2)e^{-\lambda|y|}$. Starting with $c_0 = 0$ the right hand boundary of the interval supporting the maximum rectangle within $f(y)$ satisfies

$$\max_c c \frac{\lambda}{2} e^{-\lambda c}, \tag{7.13}$$

so that $c_1 = 1/\lambda$. The area of the rectangle is $A_1 = e^{-1}/\lambda$. The next interval to the right is of the same length $c_2 - c_1$, for $c_2 = 2/\lambda$, and the area of the rectangle is $A_2 = (1/2)e^{-2}/\lambda$. Continuing, $c_3 = 3/\lambda$, $A_3 = (1/2)e^{-3}/\lambda$, and so on. By symmetry there are similar intervals and rectangles for $c_{-i} = -c_i$ and $A_{-i} = A_i$ for $i \geq 2$.

As above $q_1 = e^{-1}/(\sum_j e^{-j})$ and $q_2 = q_{-i} = e^{-i}/(2\sum_j e^{-j})$ for $i > 1$. The probability $p_1 = (1 - e^{-1})/\lambda^2$, while $p_i = p_{-i} = (e^{1-i} - e^{-i})/(2\lambda^2)$.

7.4 Main tests

In this section we describe the main testing problem in which we are interested. It has two test statistics, the ML estimator $\hat{\theta}(\mathbf{x})$ and the KL distance between the two histograms defined by Algorithm 7.1, the first with the expected bin probabilities and the second with the count ratios defined by the data. We discuss the latter test statistic and its distribution first, and then describe in detail the tests for a simple hypothesis and the composite hypothesis.

Given the density function $f(\mathbf{y}; \theta, k)$ or its best parametric representation $f(\mathbf{y}; \hat{\theta}(\mathbf{x}), k)$, depending on the hypothesis, we want to model the slowly varying part of the data, and calculate with a test statistic the degree of typicality of the data $\mathbf{x}$. We represent the slowly varying part by the numbers of times $\{n_i(\mathbf{x})\}$ the points x_t of the data $\mathbf{x} = x^n$ fall in the characteristic intervals of the hypothesis. Hence in a certain sense we consider the intervals and their counts as a quantization of the data $\mathbf{x}$. Sometimes the data $\mathbf{x}$ are not given at all, but they are replaced by the counts in a number of usually equal-bin-width groups. Clearly, there is some loss of information when only the counts are given. At any rate we have two estimates of the hypothesis, the characteristic histogram $q = q(\theta)$ or $q = q(\hat{\theta}(\mathbf{x}))$, depending on the hypothesis, and the histogram defined by the counts $\{n_i(\mathbf{x})\}$. As in estimation, the objective is to maximize a probability that depends on the two sets $\{n_i(\mathbf{x})/n\}$ and $\{q_i(\theta)\}$ for the simple hypothesis and $\{n_i(\mathbf{x})/n\}$ and $\{q_i(\hat{\theta}(\mathbf{x}))\}$ for the composite hypothesis. A large probability implies a good fit with a large degree of typicality and conversely. The probability will be defined by the KL distance $D(\{n_i/n\}\|\{q_i\})$ between the two distributions $\{n_i(\mathbf{x})/n\}$ and $\{q_i(\theta)\}$, whose minimum maximizes the probability.

Before continuing we make a digression to the traditional much studied testing problem for the shape of the density functions in the two types of hypotheses. One of the earliest test statistics, due to Pearson, is the sum of the weighted squared differences

$$Q(m-1) = \sum_{1}^{m} \frac{(n_i - np_i)^2}{np_i} = n\sum_i \frac{(n_i/n - p_i)^2}{p_i}, \quad (7.14)$$

where $\{p_i\}$ denotes the expected probabilities of the bins in a histogram estimation of the density function, defined by the hypothesis, and m denotes the number of equal-width bins. By Pearson's theorem [5] for data generated by the null hypothesis, $Q(m-1)$ has in the limit as $n \to \infty$ the χ^2 distribution of $m-1$ degrees of freedom and $m-j-1$ degrees if j of the parameters are estimated. Apart from the asymptotic nature of the approximation an immediate problem with this is that the χ^2 distribution is very flat for a large number of bins, which often is the case, and it does not become more peaked with increasing numbers of observations. Thus the decision for rejection is correct when the data are not typical for the null hypothesis, precisely when the χ^2 distribution is a bad approximation on the rejection region. Finally, the limiting χ^2 distribution does not depend on the hypothesis at all, which means that it could not be used for acceptance.

The Pearson test statistic is closely related to the other test statistic due to Neyman–Pearson and others, see for instance [11], [25], namely, the expected value of $2n \log n_i/(np_i)$ with respect to the distribution $\{n_i/n\}$, which is $2n$ times the KL distance $D(\{n_i/n\}\|\{p_i\})$. To see this we put $s_i = n_i/n$ and expand the sum $\sum_i s_i \log 1/p_i$ into Taylor series about its minimum $\sum_i s_i \log 1/s_i$ by the noise-free coding theorem, Theorem 2.1, and we get

$$\sum_i s_i \log 1/p_i = \sum_i s_i \log 1/s_i + \frac{1}{2} \sum_i \delta_i^2/s_i + \sum_i O(\delta_i^3), \quad (7.15)$$

where $p_i = n_i/n + \delta_i$. We multiply both sides by $2n$ and get

$$2nD(\{s_i\}\|\{p_i\}) = n \sum_i \frac{\delta_i^2}{p_i - \delta_i} + \sum_i O(n\delta_i^3) = Q(m-1) + \sum_i O(n\delta_i^3), \quad (7.16)$$

provided $n_i/n - p_i$ converges to zero as $n \to \infty$. Because of this the χ^2 distribution is also used as an approximation of the distribution of the KL distance test statistic although this is justified only when (7.16) converges, or when the differences $n_i/n - p_i$ converge to zero as $n \to \infty$, i.e. again the limiting χ^2 distribution is valid for wrong data for the rejection test and hence its use is not justified.

The insistence on using the χ^2 distribution is a wide spread mistake, for the KL distance divided by n is always well defined regardless of whether (7.16) converges or not, and, as we show next, it has an exact distribution, which is much better for testing than the χ^2 distribution. It retains the relevant information, which makes the KL test statistic just about as good as the optimal ML estimator.

If we temporarily regard the count ratios $s_i = n_i/n$ as the test statistics, the relevant and exact distribution is the multinomial

$$F(s;p) = \frac{n!}{n_1!n_2!\ldots n_m!}\prod_1^m p_i^{n_i}, \tag{7.17}$$

$$= \frac{n!}{n_1!n_2!\ldots n_m!} e^{-nH(s)} \times e^{-nD(s\|p)}, \tag{7.18}$$

where $p = \{p_i\}$ and $p_i = p_i(\theta)$ or $p_i = p_i(\hat{\theta}(\mathbf{x}))$ depending on whether the hypothesis is simple or complex (i.e. composite) and similarly, $s = \{s_i\}$ for $s_i = n_i(\mathbf{x})/n$, while $H(s)$ denotes the entropy. We show presently that the KL distance $nD(s\|p)$, being a scalar, defines a better test statistic with a distribution induced by the multinomial.

We write the multinomial for a simple hypothesis as $F(s(\mathbf{x}); p(\theta))$. For composite hypotheses we then replace the parameter θ by its best estimate $\hat{\theta}(\mathbf{x})$, which makes the p_is functions of $\mathbf{x}$. By Stirling's approximation of the factorials, (4.27), we have

$$\ln\frac{n!}{n_1!n_2!\ldots n_m!} - nH(s) = -\sum_{i=1}^m \ln\sqrt{n_i} + \ln\sqrt{n} - (m-1)\log\sqrt{2\pi} + \mu_n$$

$$= \frac{1}{2}\sum_i \ln\frac{n}{n_i} - (m-1)\ln\sqrt{2\pi n} + \mu_n, \tag{7.19}$$

where $\mu_n = R(n) - \sum_i R(n_i) = O(1/n)$. This approximation, except for small n, is so good that the approximation error μ_n may be ignored as we will do in the following. When in doubt we can calculate μ_n and decide if it can be ignored. Then

$$e^{-nD(s\|p)} = (2\pi n)^{(m-1)/2} F(s;p) \prod_i \sqrt{s_i}. \tag{7.20}$$

The KL distance, denoted here simply as D_s because p is kept fixed, has the distribution

$$G(D_s) = \frac{e^{-nD_s}}{\sum_t e^{-nD_t}} = e^{-nD_s}/B_p, \tag{7.21}$$

$$B_p = (2\pi n)^{(m-1)/2}\sum_t F(t;p)\prod_i \sqrt{t_i} \leq (2\pi n/m)^{(m-1)/2} m^{-1/2}, \tag{7.22}$$

where $t = \{t_i\}$ denotes the variable count ratios. To see the last inequality, either maximize $\sum_i \ln\sqrt{t_i}$ over the t_is by use of Lagrangian technique, or use the identity

$$\sum_i \ln \sqrt{t_i} = \frac{m}{2} \sum_i \frac{1}{m} \ln \sqrt{t_i} = \frac{m}{2} \left[-D\left(\left\{\frac{1}{m}\right\} \middle\| \{t_i\}\right) - \ln m \right] \leq m^{-m/2}. \tag{7.23}$$

We then see that the distribution $G(D_s)$ of the test statistic has a narrow peak exceeding $[m/(2\pi n)]^{(m-1)/2}\sqrt{m}$ at $s \approx p$, from which it shrinks at an exponential rate, where the exponent is proportional to $-n$.

From (7.20) we get the factorization

$$F(s;p) = G(D(s\|p))B_p \prod_i 1/\sqrt{s_i}, \tag{7.24}$$

written now with both of the variable symbols s and p. We see that the two distributions $F(s;p)$ and $G(D(s\|p))$ determine one another, and no information is lost in the scalar-valued test statistic $G(D(s\|p))$ from the count ratios s. Moreover, this distribution is peaked so that the typical data for the hypothesis are well separated from the atypical or random data for it. We conjecture that the KL test statistic with its distribution is either optimal or at least very close to it.

We next study the maximizing count ratios $s_i = n_i/n$ of $G(D(s\|p))$, which are rationals, or, equivalently, the minimizing ratios of $\ln 1/G(D(s\|p))$ by minimizing $\ln 1/F(s;p)$ over the reals, which is simpler. The reason is that the minimum of $\ln 1/G(D(s\|p))$ over the reals is zero and tells us nothing about the minimum over the rationals. Take the count ratios $s_i = n_i/n$ as real numbers and write the Lagrangian

$$n \sum_i \left[s_i \ln \frac{s_i}{p_i} + \frac{1}{2} \ln s_i \right] - \lambda \left(\sum_i s_i - 1 \right). \tag{7.25}$$

Putting the first derivatives to zero gives the m equations

$$n \ln(s_i/p_i) + 1/(2s_i) = \lambda - n, \ i \leq m, \tag{7.26}$$

which are to hold, together with $\sum_i s_i = 1$. Multiplying both sides by s_i and summing over i gives the equation for λ:

$$n(D+1) + m/2 = \lambda. \tag{7.27}$$

The equations to solve for the s_is such that $1 > s_i \geq 1/n$ are

$$\ln(s_i/p_i) + 1/(2ns_i) = D + m/(2n), \ i \leq m. \tag{7.28}$$

We see that for $s_i = p_i \neq O(1/n)$, $D = 0$, and both sides are $O(1/n)$, so that for any n a good start for the numerical solution is $s_i = p_i$. Denote by $\hat{s}$ the minimizing counts. Since the minimum of $D(s\|p)$ over the rationals $\{s_i\}$, say $\hat{D}$, satisfies

$$D(\hat{s}\|p) \geq \hat{D} \geq 0, \tag{7.29}$$

the excess of $D(\hat{s}\|p)$ over the zero gives an idea of the difficult to calculate $\hat{D}$, except for small n.

7.4.1 Simple hypothesis

The test of the degree of typicality of the observed data $\mathbf{x}$ relative to the hypothesis $f(\mathbf{x}; \theta, k)$ is done with the two test statistics, the ML estimate $\hat{\theta}(\mathbf{x})$ and $nD(s(\mathbf{x})\|p)$, where $s(\mathbf{x}) = \{s_i(\mathbf{x}) = n_i/n\}$, the former measuring the short term "noise" and the latter the long term drift type "noise." We saw above that the p_is can be replaced by the easier to calculate probabilities of the characteristic intervals $q_i = q_i(\theta)$ with an ignorable effect on the multinomial, which we do.

Instead of calculating the probability of the acceptance and rejection regions it will be simpler to calculate them by the ratios of the distributions of the two test statistics to their maximum, respectively, which is justified by their greatly peaked distributions

$$r(\hat{\theta}(\mathbf{x}); \theta) = \frac{g(\hat{\theta}(\mathbf{x}); \theta)}{g(\theta; \theta)}, \tag{7.30}$$

$$R(s(\mathbf{x}); \theta) = e^{-nD(s(\mathbf{x})\|q(\theta))}. \tag{7.31}$$

Moreover, in the first ratio we used the normal approximations, and in the second we took the maximum of $G(D(s\|q(\theta)))$ as unity given by the KL distance as zero.

We accept the hypothesis if both ratios satisfy the condition

$$r(\hat{\theta}(\mathbf{x}); \theta) \geq r_a, \tag{7.32}$$

$$R(s(\mathbf{x}); \theta) \geq R_a, \tag{7.33}$$

where by (6.10)

$$r_a = e^{-k/2}, \tag{7.34}$$

which requires the ML estimate to fall within the optimally estimated interval. The distribution $G(D(s\|q(\theta)))$ is for the scalar-valued test statistic, which by the CLT is also approximately gaussian. Hence, a logical choice for R_a is $R_a = e^{-1/2}$. The rejection decision could be taken on the significance level ϵ if either

$$r(\hat{\theta}(\mathbf{x}); \theta) \leq \epsilon \quad \text{or} \quad R(s(\mathbf{x}); \theta) \leq \epsilon. \tag{7.35}$$

Alternatively, both ratios could be replaced by the probabilities of the acceptance regions for the two test statistics. If $g(\hat{\theta}(\mathbf{y}); \theta)$ is taken as the normally distributed density function for the ML estimate, the acceptance region could be taken the same as above, which also by (6.18) has the probability

$$P_a = \int_{\hat{\theta} \in B_{k/n}(\theta)} g(\hat{\theta}; \theta) d\hat{\theta} = (\text{erf}(1/\sqrt{2}))^k. \tag{7.36}$$

The rejection region for the ML test statistic has the probability

$$\bar{P}_r = \int_{\hat{\theta}: r(\hat{\theta}(\mathbf{x}); \theta) \leq \epsilon} g(\hat{\theta}; \theta) d\hat{\theta}. \tag{7.37}$$

For the test with the KL statistic the acceptance and rejection probabilities for the same regions as above are given by

$$P_A = \sum_{s: R(s(\mathbf{x}); \theta) \geq R_a} G((s\|q); q), \tag{7.38}$$

$$\bar{P}_R = \sum_{s: R(s; \theta) \leq \epsilon} G(D(s\|q); q). \tag{7.39}$$

The acceptance decision is made when both test statistics fall in their acceptance region. The rejection decision is made if either test statistic falls in its rejection region. Finally, if the null hypothesis is neither rejected nor accepted we conclude that there is not sufficient statistical evidence to make the decision.

In the following example we calculate the acceptance and rejection by the ratios, which are a lot easier to compute.

Example This example is a Bernoulli trial from [5], page 421. A coin was tossed 4040 times resulting in 2048 heads and 1992 tails. The hypothesis is that the coin is balanced: $\theta = 1/2$. The ML estimate $\hat{\theta}(x^n) = 2048/4040 = 0.5069$ falls within the optimal half-interval $[0.5, 0.5 + 1/(2\sqrt{n}) = 0.5078]$. The variance of $g(\hat{\theta}; \theta)$ is $1/(4n)$, and by (7.30) $r(\hat{\theta}; \theta) = e^{-8080 \times 0.0069^2} = 0.68$, which exceeds the same ratio $r_a = e^{-1/2} = 0.607$ at the boundary of the acceptance region. The KL distance test statistic in (7.31) is just about unity, implying the acceptance.

98 **Hypothesis testing**

This test result is much more convincing than the "non-rejection" decision in [5], which is based on the χ^2 distribution, even though the approximation in this case is good, because the sample size 4040 is large and the degree of freedom of the χ^2 distribution just 1. In Section 7.4.3 we discuss a much worse application of χ^2 tests on basically an estimation problem as a model selection.

7.4.2 Composite hypothesis

The test of the composite hypothesis, of which we only specify that it belongs to a subset of the parameter space, is similar but more difficult, especially when the entire parameter is to be estimated. This is because both the rejection and acceptance must be based on the probability calculated only from the characteristic histogram. The hypothesis is the optimal model $f(\mathbf{y}; \hat{\theta}(\mathbf{x}), k)$, which obviously cannot be rejected or accepted without further evidence. To test for its shape we calculate the KL distance conditioned on the probabilities of the characteristic intervals or the counts q_i of the density function $f(\mathbf{y}; \hat{\theta}(\mathbf{x}), k)$ as in (7.31) for $\theta = \hat{\theta}(\mathbf{x})$. These together with the counts determined by the data $s(\mathbf{x}) = \{s_i(\mathbf{x})\}$ define the KL distance and the test statistic.

An important special case of composite hypotheses is when a part of the parameter of models in a class, say $\mathcal{M}_k$, is fixed and only the rest is estimated. For $\theta = (\alpha, \beta)$, the first part α is fixed, often as zero, and the composite hypothesis $f(\mathbf{y}; \hat{\theta}(\mathbf{x}))$ is defined by the parameter $\hat{\theta}(\mathbf{x}) = (\alpha, \hat{\beta}(x^n))$, where $\hat{\beta}$ is defined by

$$\max_{\beta} f(\mathbf{x}; (\alpha, \beta)). \tag{7.40}$$

The first test statistic for α is the first part $\hat{\alpha}(\mathbf{x})$ of the ML estimate of the unrestricted parameter $(\hat{\alpha}(\mathbf{x}), \hat{\beta}(\mathbf{x}))$, or

$$\max_{\alpha} f(\mathbf{x}; (\alpha, \hat{\beta}(\mathbf{x}))). \tag{7.41}$$

Its distribution is defined by

$$g(\hat{\alpha}; (\alpha, \hat{\beta}(\mathbf{x}))) = \int_{\mathbf{y}: \hat{\alpha}(\mathbf{y})=\hat{\alpha}} f(\mathbf{y}; (\hat{\alpha}, \hat{\beta}(\mathbf{x}))) d\mathbf{y}. \tag{7.42}$$

The second test statistic is the KL distance $D(s(\mathbf{x})\|q(\mathbf{x}))$, where q is defined by the characteristic intervals of the density function $f(\mathbf{x}; (\alpha, \hat{\beta}(\mathbf{x})))$, and s is defined by the counts $s_i = n_i/n$ in these intervals. Its distribution is defined by $G(s, q)$ in (7.21).

For the test write $\theta = (\alpha, \hat{\beta}(\mathbf{x}))$, and replace (7.30) and (7.31) by

$$r(\hat{\alpha}(\mathbf{x}); \theta) = \frac{g(\hat{\alpha}(\mathbf{x}); \theta)}{g(\alpha; \theta)}, \qquad (7.43)$$

$$R(s(\mathbf{x}); \theta) = e^{-nD(s(\mathbf{x}) \| q(\theta))}. \qquad (7.44)$$

We accept the hypothesis, just as in the case with simple hypotheses, if both ratios satisfy the condition

$$r(\hat{\alpha}(\mathbf{x}); \theta) \geq r_a, \qquad (7.45)$$

$$R(s(\mathbf{x}); \theta) \geq R_a, \qquad (7.46)$$

where for k' denoting the number of components in α

$$r_a = e^{-k'/2}, \qquad (7.47)$$

and $R_a = e^{-1/2}$.

The rejection decision is taken on the significance level ϵ if either

$$r(\hat{\alpha}(\mathbf{x}); \theta) \leq \epsilon \text{ or } R(s(\mathbf{x}); \theta) \leq \epsilon. \qquad (7.48)$$

In the case where no fixed parameters exist we calculate the ratio (7.31)

$$R(s(\mathbf{x}); \hat{\theta}(\mathbf{x})) = e^{-nD(s(\mathbf{x}) \| q(\hat{\theta}(\mathbf{x})))} \qquad (7.49)$$

with acceptance or rejection decided by whether this ratio satisfies the inequalities in (7.33) and (7.35), respectively.

We next consider two examples, in both of which we use the ratio calculations.

Example In this first example we consider the test for the normality of the mean June temperatures during 1841–1940 in Table 30.4.2 in [5] and partly duplicated in Table 7.1. The second column gives the number of days in the temperature intervals in the first column, and the third column gives the mean temperatures for the given days calculated from the 100 observed temperature values, which are not given here. Notice that the mean temperatures do not fall within the temperature intervals.

The test in the cited reference was done using the approximate χ^2-test statistic. There is a bit of a problem in calculating the mean and the standard deviation because instead of the 100 temperatures values only their numbers in ten intervals

Table 7.1. *Data from Table 30.4.2 in [5]*

degrees Celsius	number of days observed	mean temperatures
12.5–12.9	12	7.89
–12.4	10	12.89
13.0–13.4	9	10.20
13.5–13.9	10	11.93
14.0–14.4	19	12.62
14.5–14.9	10	12.08
15.0–15.4	9	10.46
15.5–15.9	6	8.19
16.0–16.4	7	5.81
16.5–	8	7.98

were given. The mean value of the temperature x estimated from the occurrence counts is $\bar{x} = 14.23$. Instead of estimating the excess and skewness as in [5], we just estimate the standard deviation needed for the ML estimate from the interval boundaries and the counts in the second column. We partition first the compact range $(12.5, 16.5)$ into eight intervals of equal length, i.e. 0.5 degrees Celsius, and assume that the two tails are also 0.5 degrees Celsius each. By calculating the boundaries from the mean value, marked at 0, we get the boundaries of the intervals as $-2.23, -1.73, -1.23, -0.73, -0.23, 0.27, 0.77, 1.27, 1.77, 2.27$, and their counts $10, 12, 9, 10, 19, 10, 9, 6, 7, 8$. From the given counts in each we then calculate a histogram, where at each of the five temperatures we place a fractional count, the given count divided by 5. For instance, from the interval with 19 counts we calculate the contribution to the variance as follows $\frac{19}{5}[0.2^2 + 0.1^2 + 0.1^2 + 0.2^2] = 0.38$. In this way we get the variance as 0.01×212, and the standard deviation as $s = 1.46$, which gives the (approximately) optimal values of the parameters defining the null hypothesis, the normal (0,2.12) density function.

Instead of normalizing the null hypothesis we transform the characteristic intervals of the normalized (0,1) density function. They are just σ times the ones in (7.10), or $c_1 = -2.36$, $c_2 = -1.46$, $c_3 = 1.46$, and $c_4 = 2.36$, which cover the entire range. The corresponding q values are also σ times those in (7.11), which after normalization give $q_1 = q_3 = 0.205$ and $q_2 = 0.59$. The counts of the characteristic intervals are $22, 57$, and 21. Together with the probabilities q we get the KL distance in natural logarithm units or nats $nD(\{n_i/n\}\|q) = 1$. Further, by (7.31) $e^{-nD} = 0.367 < e^{-1/2}$, which does not make the hypothesis supported by the data. In [5] the χ^2-test statistic gave a satisfactory agreement with the null hypothesis and non-rejection.

Example This second example is taken from the data in Table 31.3.7 in [5] where we have the actual numbers of the observed data rather than just their groups. Ten people each tried two types of sleeping pills, A and B. The differences z between the number of hours of sleep gained with A and that with B were $1.2, 2.4, 1.3, 1.3, 0, 1.0, 1.8, 0.8, 4.6, 1.4$. Hence, for instance, the first person reported that pill A helped him or her to sleep 1.2 hours more than after having taken pill B. The (null) hypothesis is that z is normally distributed with mean zero and some variance so that the composite hypothesis refers to a subset of the normally distributed models. Actually, it seems quite obvious that pill A is better, and any reasonable test should show it.

The ML estimates of the mean and the standard deviation are $\hat{z} = 1.58$, $\hat{\sigma} = 1.23$. Instead of the exact density function for the mean estimates under the hypothesis, we use the normal approximation (6.10)

$$g(\hat{\mu}; (0, \hat{\sigma}^2)) = \frac{\sqrt{10}/1.23}{\sqrt{2\pi}} e^{-10\hat{\mu}^2/(2 \times 1.23^2)} \tag{7.50}$$

to calculate the ratio (7.30)

$$r_a = e^{-10 \times 1.58^2/(2 \times 1.23^2)} = 0.00026. \tag{7.51}$$

This means rejection with overwhelming statistical significance.

In [5] the zero-mean test was done with the test statistic $t = \sqrt{9}(\hat{z} - 0)/s$, which has Student's distribution with 9 degrees of freedom, provided the null hypothesis is true. The data give the value $t = 4.064$, which falls inside the tail of probability 0.01, indicating rejection.

We next test the hypothesis that the data have been generated by sampling an optimally estimated normal distribution $(\hat{z}, \hat{\sigma}^2)$. The normal $(\hat{z}, \hat{\sigma}^2)$ density function defines the characteristic intervals, only three of which, $\{(-1.23, 1.23), (1.23, 2), (2.5, 2.85)\}$, have positive counts, $8, 1, 1$, which define $s = 0.1, 0.8, 0.1$. The distribution $p = \{0.04, 0.1, 0.68, 0.1, 0.04\}$, (7.12), defines the corresponding three probabilities $q = p = \{0.14, 0.68, 0.14\}$. By (7.31) we get

$nD(s\|q) = 0.627$ and $r_2 = 2^{-0.627} = 0.53$, which indicates acceptance of the hypothesis.

Finally, since the data are non-negative we test the exponential hypothesis $p(x) = \lambda e^{-\lambda x}$. We get $\hat{\lambda}(x^{10}) = 10/15.8 = 0.63$. The three characteristic intervals that cover the data are $[0, 1.58)$, $[1.58, 3.16)$, and $[3.16, 4.74)$, which give the counts $s = \{7, 2, 1\}$. The corresponding areas of the rectangles are 0.58, 0.21, and 0.079, which after normalization give $q_1 = 0.667$, $q_2 = 0.241$, and $q_3 = 0.09$. These give the KL distance $nD = 0.034$, and by (7.31) $r_2 = e^{-nD} = 0.967$, which implies acceptance of the hypothesis with a better match than the previous normal model.

7.4.3 Likelihood ratio

We conclude this chapter with a brief discussion of another type of application of χ^2 testing to the composite hypothesis where some of the parameters are fixed; see, for instance, chapter 22 in [11]. Let $\mathcal{M}_k = \{f(x^n; \theta) : \theta \in \Theta\}$, where Θ is a subset of R^k. The null hypothesis Θ_0 is a subset of Θ such that the first r of the parameters $\theta \in \Theta$ are zero or fixed. In the notation above this is the part α. The test statistic is the logarithm of the likelihood ratio

$$\lambda_n = \frac{\sup_{\theta \in \Theta_0} \prod f(x_j; \theta)}{\sup_{\theta \in \Theta} \prod f(x_j; \theta)}, \tag{7.52}$$

and the distribution of $-2 \log \lambda_n \to \chi_r^2$ in law, provided the data are generated by some model in Θ_0.

The trouble with this is that for the rejection decision we need a good approximation of the likelihood ratio for data that are not generated by the null hypothesis, i.e. where the first r parameters are not small. But the χ_r^2 distribution for the likelihood ratio is a reasonable approximation only when all the parameters are small and the trouble with this is that under the hypothesis the test statistic is n times a sum of k numbers all converging to zero, no matter how large n is. This means that it is very hard to separate the r and $k - r$ numbers that isolate the null hypothesis even if the distribution of the test statistic is the χ_r^2 distribution, which is the case only to an approximation. Furthermore, for the χ_r^2 approximation to be even reasonable the convergence takes place for the wrong data, namely, where the first r parameters are not small.

7.4 Main tests

The same test is sometimes suggested for the estimation of the number of parameters, $k - r$. In denoising problems, discussed in the next chapter, which are regression problems, both of the numbers k and r are typically in the hundreds or thousands, and to apply a χ^2 distribution with so many degrees of freedom, whether accurate or not, is not going to tell us anything worthwhile about the null hypothesis.

A better way to estimate the parameters $\theta_{r+1}, \ldots, \theta_{r+\nu}$, including their number $\nu = k - r$, or more generally, $\theta_{(1)}, \ldots, \theta_{(\nu)}$, wherever they occur, is done by maximization of the density function (4.49):

$$\max_{\nu} f(x^n; \hat{\theta}(x^n), \nu)/\hat{C}_{\nu}, \tag{7.53}$$

where $\hat{C}_{\nu} = \int g(\hat{\theta}; \hat{\theta}, \nu) d\hat{\theta}$, and $g(\hat{\theta}; \theta, \nu)$ is the density function on the ML estimates, induced by $f(x^n; \theta, \nu)$. This avoids arbitrary levels of tests and flat distributions for the test statistics, and, in fact, is actually optimal in separating the first r zero parameters from the relevant parameters of the null hypothesis, which appears to be a needlessly convoluted way to formulate an estimation problem of the kind discussed in Chapter 3.

8 Denoising

The important so-called denoising problem is to break up a data sequence $x^n = x_1, x_2, \ldots, x_n$, written as a row vector, as follows

$$x^n = \hat{x}^n + e^n, \qquad (8.1)$$

where e^n represents "noise" and $\hat{x}^n$ the denoised part of the observed sequence x^n as the cleaned signal. This can obviously be done in several ways depending on what we mean by noise, which in most solutions to the problem is actually undefined and hence effectively constitutes the part in the sequence that is removed. Intuitively, we associate with "noise" small rapid fluctuations, which when removed leave a "smooth" slowly varying denoised signal $\hat{x}^n$, which has the intended information in the data x^n. As even the denoised signal can be complex and "random" looking we want to fit probability models to the data, and the only way to separate the two parts is to model them differently.

An excellent way to formalize and measure the degrees of "smoothness" and rapidity of fluctuations is by a linear wavelet transform: $x^n \mapsto x^n \mathcal{W} = c^n$, which like all linear transforms can be represented by an $n \times n$ matrix, also denoted by $\mathcal{W}$. In denoising the matrix is very large, typically $n = 1000$ or more, which conveniently is defined by a wavelet transform algorithm and need not even be written down as a matrix. Especially convenient are wavelet transforms for which the matrix is orthonormal, its inverse given by the transpose $\mathcal{W}'$; see, for instance, [8]. Then $c^n = x^n \mathcal{W}$ represents the coefficients when x^n is expressed as a linear combination of orthonormal basis vectors, the columns of $\mathcal{W}'$. These define an orthonormal basis for the signals, and they range from slowly varying to rapidly varying functions, analogous to the sinusoidal bases of Fourier series.

Geometrically the denoising problem amounts to partitioning the n basis vectors into two subsets, defined by their indices $\gamma = \{i_1, i_2, \ldots, i_k\}$ and the rest $\bar{\gamma}$ in such a manner that when $\hat{x}^n$ and e^n are expressed as a linear combination of the basis vectors with indices in γ and $\bar{\gamma}$, respectively, the sum of their squared lengths is minimum. This amounts to minimizing the code length if we model the two signals by gaussian density functions.

8.1 Hard thresholding

The denoising problem can be formally studied within the estimation theory, in which "noise" is defined as the part of the data which cannot be compressed with a chosen model class, so that in our terminology it is random relative to the model class. Consider a linear quadratic regression problem, defined by the matrix $\mathcal{W}$ with its transpose $\mathcal{W}'$ as the inverse

$$x^n = \beta^n \mathcal{W}' + e^n = \hat{x}^n + e^n, \qquad (8.2)$$
$$\beta_i \neq 0 \text{ for } i \in \gamma, \qquad (8.3)$$
$$\beta_i = 0 \text{ for } i \in \bar{\gamma}, \qquad (8.4)$$

where $\gamma = \{i_1, i_2, \ldots, i_k\}$ consists of indices defining the structure. We denote by β the k-component row vector consisting of the non-zero components of the n-component row vector β^n. We also write $\theta = (\beta, \tau)$, and when the superindex in x^n is not needed we write $x^n = \mathbf{x}$. Finally, we assume $0 < k < n$. Otherwise the denoising problem is trivial, because the data consist of either all noise or no noise, and the separation is simply a matter of semantics.

We model e^n as an iid zero-mean gaussian sequence $\phi(e_i; 0, \tau)$ of variance τ, which induces distributions both for the signals $\hat{x}^n$ and x^n, and we get a class of models of type $\mathcal{M}_\gamma$, in which hyperparameters are needed to keep the normalizing integral finite. This so-called MDL denoising solution was used in [45], which was inspired by an earlier work, [23]. We describe a simplified version of it, which, moreover, has no hyperparameters, for they are optimized.

The normal models $\phi(e_i; 0, \tau)$ for the components e_i induce a normal density function $f(x^n; \gamma, \beta, \tau) = \prod_i \phi(x_i; \hat{x}_i, \tau)$, where $\hat{x}_i$ is the ith component of $\beta^n \mathcal{W}'$

in (8.2). We have the well-known important factorization:

$$f(y^n; \beta, \tau, \gamma,) = f(y^n | \hat{\beta}, \hat{\tau}, \gamma,) p_1(\hat{\beta}; \beta, \tau) p_2(n\hat{\tau}/\tau; \tau) \frac{n}{\tau}, \tag{8.5}$$

$$f(y^n | \hat{\beta}, \hat{\tau}, \gamma,) = (2\pi)^{\frac{k-n}{2}} n^{-n/2} \Gamma(\frac{n-k}{2}) 2^{\frac{n-k}{2}} \hat{\tau}^{1-\frac{n-k}{2}}, \tag{8.6}$$

$$p_1(\hat{\beta}; \beta, \tau) = \frac{n^{k/2}}{(2\pi\tau)^{k/2}} e^{-\frac{n}{2\tau}(\hat{\beta}-\beta)'(\hat{\beta}-\beta)}, \tag{8.7}$$

$$p_2(n\hat{\tau}/\tau; \tau) = (n\hat{\tau}/\tau)^{\frac{n-k}{2}-1} e^{-n\frac{\hat{\tau}}{2\tau}} 2^{-\frac{n-k}{2}} / \Gamma\left(\frac{n-k}{2}\right). \tag{8.8}$$

The factor p_1 is normal with mean β and covariance $(\tau/n)I_k$, where I_k is the $k \times k$ identity matrix, while p_2 is the χ^2 distribution for $n\hat{\tau}/\tau$ with $n-k$ degrees of freedom. Moreover, the two factors are independent, and the statistic $\hat{\theta} = (\hat{\beta}, \hat{\tau})$ is said to be sufficient. This is because $f(y^n | \hat{\theta}(y^n); \gamma) = h(y^n)$ depends only on the data and not on θ.

Since $f(y^n; \theta, \gamma) = f(y^n, \hat{\theta}(y^n); \theta, \gamma)$ we can write it as the product of the marginal density of $\hat{\theta}$

$$p(\hat{\theta}; \theta, \gamma) = p_1(\hat{\beta}; \theta) p_2(n\hat{\tau}/\tau; \tau) \frac{n}{\tau} \tag{8.9}$$

and the conditional density of y^n given $\hat{\theta}$

$$f(y^n; \theta, \gamma) = f(y^n | \hat{\theta}(y^n); \theta, \gamma) p(\hat{\theta}(y^n); \theta, \gamma). \tag{8.10}$$

The least squares estimates of β_i for $i \in \gamma$ are given by $\hat{\beta}_i(x^n) = c_i(x^n)$ and 0 for other indices, where $c^n = x^n \mathcal{W} = c_1, c_2, \ldots, c_n$, which further give the ML estimates of the variance

$$\hat{\tau}(x^n) = \frac{1}{n} \sum_{1}^{n} (x_i - \hat{x}_i)^2. \tag{8.11}$$

With (8.5) and the subsequent equations the maximized likelihood at $\mathbf{y} = y^n$ is given by

$$f(\mathbf{y}; \hat{\beta}(\mathbf{y}), \hat{\tau}(\mathbf{y}), \gamma) = A_{n,k} f(\mathbf{y} | \hat{\beta}(\mathbf{y}), \hat{\tau}(\mathbf{y}), \gamma) \hat{\tau}(\mathbf{y})^{-k/2-1} \tag{8.12}$$

$$A_{n,k} = \left(\frac{n}{2e}\right)^{n/2} \frac{1}{\pi^{k/2} \Gamma\left(\frac{n-k}{2}\right)}, \tag{8.13}$$

which does not depend on $\hat{\beta}(\mathbf{y})$. To get the capacity this will be integrated over $Y(\tau_0, R) = \{\mathbf{y} : \hat{\beta}(\mathbf{y})\hat{\beta}'(\mathbf{y}) \leq R, \hat{\tau}(\mathbf{y}) \geq \tau_0\}$, where R and τ_0 are hyperparameters. As we shall see $n\tau_0$ sets a lower bound for the sum of the squares $\sum_i e_i^2$, while nR sets an upper bound for the sum of the squares of the projection $\hat{x}^n$, and because of the symmetry either one can be regarded as noise. Typical data for denoising is such that the noise is smaller than the meaningful signal, the projection, and we set $\tau_0 < R$.

Integrating the conditional $f(y^n|\hat{\theta}(y^n);\theta,\gamma) = h(y^n)$ over y^n such that $\hat{\theta}(y^n)$ equals any fixed value $\hat{\theta}$ yields unity. The integral for the normalizing coefficient becomes

$$\hat{C}_\gamma(R,\tau_0) = A_{n,k} \int_{\tau_0}^{\infty} \hat{\tau}^{-k/2-1} d\hat{\tau} \int_{\tau_0 \leq \hat{\beta}'\hat{\beta} \leq R} d\hat{\beta} \qquad (8.14)$$

$$= \left(\frac{n}{2e}\right)^{n/2} \times \frac{4}{k^2} \times \frac{(R/\tau_0)^{k/2}}{\Gamma(\frac{n-k}{2})\Gamma(\frac{k}{2})}, \qquad (8.15)$$

which gives the normalized density function as

$$\ln 1/\hat{f}(\mathbf{x};\tau_0,R,\gamma) = \frac{n}{2}\ln \hat{\tau}(\mathbf{x}) + \ln \hat{C}_\gamma(R,\tau_0), \qquad (8.16)$$

where $\mathbf{x}$ is required to fall within $Y(\tau_0, R)$.

To optimize the hyperparameters $\tau_0 < R$ notice in (8.15) that the smallest ratio R/τ_0 such that $\mathbf{x} \in Y(\tau_0, R)$, which minimizes (8.16), is $\hat{R}(\mathbf{x})/\hat{\tau}(\mathbf{x})$. Indeed, $\hat{R}(\mathbf{x})/\hat{\tau}(\mathbf{x}) \leq R/\hat{\tau}(\mathbf{x}) \leq R/\tau_0$. With these we obtain

$$\hat{f}(\mathbf{y};\gamma,\hat{\tau}(\mathbf{x}),\hat{R}(\mathbf{x})) = \frac{\hat{\tau}(\mathbf{y})^{-n/2}}{\hat{C}_\gamma(\hat{R}(\mathbf{x}),\hat{\tau}(\mathbf{x}))}. \qquad (8.17)$$

The next step is to calculate

$$\hat{f}(\mathbf{x};\gamma) = \frac{\hat{f}(\mathbf{x};\gamma,\hat{\tau}(\mathbf{x}),\hat{R}(\mathbf{x}))}{\hat{C}_\gamma}, \qquad (8.18)$$

where $\hat{C}_\gamma$ is given by

$$\hat{C}_\gamma = \int_{Y(\hat{\tau}(\mathbf{x}),\hat{R}(\mathbf{x}))} \hat{f}(\mathbf{y};\gamma,\hat{\tau}(\mathbf{y}),\hat{R}(\mathbf{y})) d\mathbf{y}. \qquad (8.19)$$

By (8.9) and (8.10) we have the factorization

$$\hat{f}(\mathbf{y};\gamma,\hat{\tau}(\mathbf{y}),\hat{R}(\mathbf{y})) = \alpha_n \hat{f}(\mathbf{y}|\gamma,\hat{\tau}(\mathbf{y}),\hat{\beta}(\mathbf{y}))\frac{k}{2}\Gamma(k/2)^{-1}\hat{R}(\mathbf{y})^{-k/2}\hat{\tau}(\mathbf{y})^{-1}, \qquad (8.20)$$

where α_n depends only on n. As in (8.15) we can integrate the conditional while keeping $\hat{\beta}(\mathbf{y}) = \beta$ and $\hat{\tau}(\mathbf{y}) = \tau$ constant, which gives unity. Then we integrate the resulting function of τ over the range $B = [\hat{\tau}(\mathbf{x}), \hat{R}(\mathbf{x})]$ and $\hat{R}(\mathbf{y})^{-k/2}$ over the same range assuming $\hat{R}(\mathbf{x}) > \hat{\tau}(\mathbf{x})$. We get

$$\hat{C}_\gamma = \alpha_n \frac{k}{2} \Gamma(k/2)^{-1} \hat{R}(\mathbf{x})^{-k/2} \int_{\tau \in B} \tau^{-1} d\tau \int_{R \leq \hat{R}(\mathbf{x})} dR \int_{\beta'(\mathbf{y})\beta(\mathbf{y}) = R} d\mathbf{y}$$
$$= \alpha_n \frac{k^2}{4} [\ln \hat{R}(\mathbf{x}) / \ln \hat{\tau}(\mathbf{x})]^2, \tag{8.21}$$

where we used the fact that

$$\int_{\mathbf{y}: \beta'(\mathbf{y})\beta(\mathbf{y})=R} d\mathbf{y} = \frac{k\pi^{k/2}}{2\Gamma(k/2)} R^{k/2-1} \tag{8.22}$$

is the surface area of the hyperball of radius R.

Application of Stirling's approximation to the gamma functions gives

$$\ln 1/\hat{f}(\mathbf{x}; \gamma) = \frac{n-k}{2} \ln \frac{S - S_\gamma}{n-k} + \frac{k}{2} \ln \frac{S_\gamma}{k} + \frac{1}{2} \ln(k(n-k))$$
$$+ 2 \ln \ln \frac{\hat{R}(\mathbf{x})}{\hat{\tau}(\mathbf{x})} + \ln \alpha_n, \tag{8.23}$$

the constant depending only on the fixed n. This differs from the corresponding formula in [45] by the next to last term, which replaces a term involving four hyperparameters.

To optimize γ for a fixed k notice first that for any γ

$$n\hat{R}(\mathbf{x}) = \sum_{i \in \gamma} c_i^2 = S_\gamma, \tag{8.24}$$

$$n\hat{\tau}(\mathbf{x}) = \sum_{i \in \bar{\gamma}} c_i^2 = S_{\bar{\gamma}}. \tag{8.25}$$

To see the second equality we have first $\mathbf{x} = \hat{\mathbf{x}} + \mathbf{e}$, taken as row vectors, and then

$$\mathbf{xx}' = \hat{\mathbf{x}}\hat{\mathbf{x}}' + (\mathbf{x} - \hat{\mathbf{x}})(\mathbf{x} - \hat{\mathbf{x}})' \tag{8.26}$$

by the fact that $\hat{\mathbf{x}}$ is orthogonal to $\mathbf{e}$. The second term in the sum is $n\hat{\tau}(\mathbf{x})$. By postmultiplying the terms in the sum with the inverse $\mathcal{W}'$ we have $\mathbf{cc}' = \sum_{i \in \gamma} c_i^2 + n\hat{\tau}(\mathbf{x})$.

It was shown in [45] that the optimizing $\hat{\gamma}$ of size k is given by the indices of the k largest or smallest squared coefficients for some k. For convenience we give the derivation. Differentiate (8.23) with respect to $u_i = c_i^2$ for $i \in \gamma$, which gives

$$\frac{k}{S_\gamma} - \frac{n-k}{\mathbf{cc}' - S_\gamma}. \tag{8.27}$$

The derivative is positive if and only if $S_\gamma/k < S_{\bar\gamma}/(n-k)$, while the second derivative is always negative, which means that (8.23) as a function of u_i is concave.

If for some γ, $S_\gamma/k > S_{\bar\gamma}/(n-k)$, we can reduce (8.23) by replacing the smallest c_i^2 for $i \in \gamma$ by a larger square for $i \in \bar\gamma$, which increases S_γ/k and gives a better γ. Continuing in this way we get $\gamma = \{(1),(2),\ldots,(k)\}$ as the set of the indices of the k largest squares. If for some γ, $S_\gamma/k < S_{\bar\gamma}/(n-k)$, the same process will deliver a set of the indices of k smallest squared coefficients, and the optimal set is the one for which (8.23) is smaller.

If we introduce $\hat\gamma(\hat k(\mathbf{x}))$ into the density function $\hat f(\mathbf{x};\gamma)$ which maximizes it, we do not get a density function. To get one we either normalize it to give $\hat f(\mathbf{x})$ or we add the code length for $\hat\gamma(\hat k(\mathbf{x}))$ to (8.23). By Theorem 4.1 the normalization would be unnecessary if we knew that there were no common factors in $\hat f(\mathbf{x})$, which we do not know without calculating the normalizing coefficient. It is much easier to calculate the two-part code length

$$\min_k \left[\ln 1/(\hat f(\mathbf{x});\hat\gamma(k)) + \ln \binom{n}{k}\right]. \tag{8.28}$$

Because $c(\mathbf{x}) \leftrightarrow \mathbf{x}$, the density function defined by (8.28) is complete. By Stirling's approximation (4.28) we get the final criterion

$$\min_k \left[\frac{n-k}{2}\ln\frac{S-S_{\hat\gamma(k)}}{n-k} + \frac{k}{2}\ln\frac{S_{\hat\gamma(k)}}{k} + nh(k/n)\right], \tag{8.29}$$

which agrees with that in [47] despite the different format.

Once the optimal set $\hat\gamma(\hat k)$ is found in at most $O(2n)$ calculations of (8.29) we need to calculate the inverse wavelet transform $\hat x^n = c^n(\hat\gamma)\mathcal{W}$ to get the denoised signal, where $c^n(\hat\gamma)$ is defined by the coefficients c_i for $i \in \hat\gamma$ and 0 otherwise.

8.2 Soft thresholding

The numerous ad hoc algorithms in the literature actually have produced a useful variant: instead of modeling the denoised signal with the retained coefficients c_i for $i \in \gamma$ while modeling the noise by setting the coefficients $c_i = 0$ for $i \in \bar\gamma$, which amounts to "hard" thresholding, one may construct the model so that the coefficients in $\bar\gamma$ are not set to zero. Instead they are left small and made gradually to increase to full size. This poses a bit of a problem for MDL denoising, because

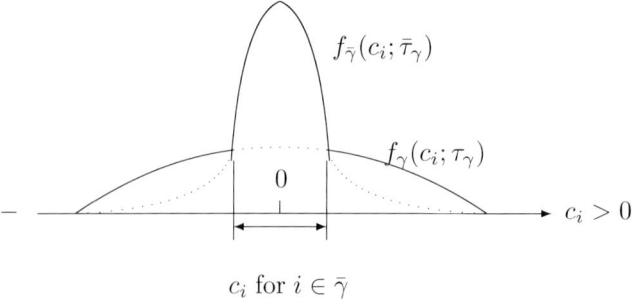

Figure 8.1 Functions for soft thresholding.

in the model class of type $\mathcal{M}_k$ each model must have k non-zero parameters while the rest are considered zero. How then can we model the denoised signal $\hat{x}^n$?

In [47] the model (8.23) was shown to consist of the product of two gaussian models, one for the coefficients in γ and the other for the coefficients in $\bar{\gamma}$:

$$f(c^n; 0, \tau_\gamma, \tau_{\bar{\gamma}}) = \prod_{i \in \gamma} \phi(c_i; 0, \tau_\gamma) \prod_{i \in \bar{\gamma}} \phi(c_i; 0, \tau_{\bar{\gamma}}), \qquad (8.30)$$

where $\tau_\gamma = S_\gamma/k$ and $\tau_{\bar{\gamma}} = S_{\bar{\gamma}}/(n-k)$. There is a restriction for data: it must be the case that $S_\gamma > S_{\bar{\gamma}}$, which amounts to adding the requirement $\hat{R}(\mathbf{x}) > \hat{\tau}(\mathbf{x})$ to the optimized hyperparameters, i.e. $\hat{\mathbf{x}}\hat{\mathbf{x}}' > \mathbf{e}\mathbf{e}'$.

In the cited reference an efficient soft thresholding was achieved in a somewhat complex manner. We describe a simpler albeit untested way. We extend both of the gaussians to all the coefficients

$$f_\gamma(c^n; \tau_\gamma) = \prod_1^n \phi(c_i; 0, \tau_\gamma),$$

$$f_{\bar{\gamma}}(c^n; \tau_{\bar{\gamma}}) = \prod_1^n \phi(c_i; 0, \tau_{\bar{\gamma}}),$$

see Figure 8.1, and define

$$w(c_i) = \frac{\min\{f_\gamma(c_i; \tau_\gamma), f_{\bar{\gamma}}(c_i; \tau_{\bar{\gamma}})\}}{f_\gamma(c_i; \tau_\gamma) + f_{\bar{\gamma}}(c_i; \tau_{\bar{\gamma}})}, \quad i \in \gamma, \qquad (8.31)$$

$$w(c_i) = \frac{\max\{f_\gamma(c_i; \tau_\gamma), f_{\bar{\gamma}}(c_i; \tau_{\bar{\gamma}})\}}{f_\gamma(c_i; \tau_\gamma) + f_{\bar{\gamma}}(c_i; \tau_{\bar{\gamma}})}, \quad i \in \bar{\gamma}. \qquad (8.32)$$

Now we define a one-to-one transformation of the sequence c^n:

$$\tilde{c}_i = w(c_i)c_i, \qquad (8.33)$$

which can be encoded with $f_\gamma(\tilde{c}_i; \tau_\gamma)$ for $i \in \gamma$ and with $f_{\bar{\gamma}}(\tilde{c}_i; \tau_{\bar{\gamma}})$ for $i \in \bar{\gamma}$. Notice that $w(c_i)$ is one half on the boundary of γ and $\bar{\gamma}$. It grows to unity for $|c_i|$ large but remains less than unity for $|c_i|$ small.

Because no coefficients are set to zero we need to redefine the set of indices γ, which is done as follows. Put

$$\gamma = \{i : f_\gamma(c_i; \tau_\gamma) \geq f_{\bar{\gamma}}(c_i; \tau_{\bar{\gamma}})\}. \tag{8.34}$$

The optimal $\hat{\gamma}$ still consists of the largest squared coefficients for the number k that minimizes (8.23). The denoised signal is obtained by the inverse transformation $\hat{x}^n = \tilde{c}^n \mathcal{W}'$.

In conclusion, a detailed investigation shows that this soft denoising scheme is actually closely related to that in [47], and since that works very well so should the simpler one described here.

9 Sequential models

Sequential models differ from the batch models in the classes $\mathcal{M}_k$ and $\mathcal{M}$ discussed up to now, in that they define random processes. They also have the advantage that their maximum capacity is generally easier to calculate, because it involves integrals or sums over one variable y_t at a time rather than the ML parameters in the batch mode, which have to be integrated over all sequences. Further, the sequentially maximized likelihood is in general greater than the maximized likelihood of batch models. To simplify the notation we show the number of parameters only when needed.

The ML estimates are calculated for each symbol in an ordered string $x^n = x_1, x_2, \ldots, x_n$, and the conditional

$$f(x_{t+1}|x^t; \hat{\theta}(x^{t+1})) = f(x^{t+1}; \hat{\theta}(x^{t+1}))/f(x^t; \hat{\theta}(x^{t+1})) \qquad (9.1)$$

is normalized

$$\hat{f}(x_{t+1}|x^t) = f(x_{t+1}|x^t, \hat{\theta}(x^{t+1}))/K(x^t), \qquad (9.2)$$

$$K(x^t) = \int f(y|x^t; \hat{\theta}(x^t, y))dy, \qquad (9.3)$$

where $x^t, y = x_1, x_2, \ldots, x_t, y$, to give the density function

$$\hat{f}_S(x^n) = \hat{f}(x^m) \prod_{t=m}^{n-1} \hat{f}(x_{t+1}|x^t). \qquad (9.4)$$

The initial segment is the normalized batch model $\hat{f}(x^m) = \hat{f}(x^m; k)$, which is needed, because in order to calculate the ML estimate of k components a minimum number $m_k = m(x^n)$ of the initial data points are needed. The capacity is given by

$$\log \hat{K}(n) = \log \hat{C}_m + \sum_{t=m+1}^{n} \log K(x^t), \qquad (9.5)$$

where to simplify matters we restrict the strings such that $m(y^n)$ equals the constant $m(x^n) = m$. The number m can be optimized:

$$\max_{m : m \geq m(x^n)} \hat{f}_S(x^n). \tag{9.6}$$

A related earlier sequential model is the predictive MDL model, also suggested by Dawid, [9], as a "prequential" model, where the parameters in the conditional are calculated from the past data $f(x_{t+1}|x^t, \hat{\theta}(x^t))$ that do not include x_{t+1}, and no normalization is needed, because the coefficient is unity.

We show next that the best sequential model is (9.4), where the ML estimates are calculated at each time instance rather than at some other set of instances. Let then $T_\nu = \{t_1 = m < t_2 < \cdots < t_\nu = n\}$ be a set of integers. A special case is the batch sequence $m = n$.

Since we do not need to indicate the individual components of the ML estimates we use the simpler notation $\hat{\theta}_{t_i}$ for the entire k-dimensional vector estimate, the lower index indicating the quantity of data t_i from which the estimate is made. Let

$$f(x^n; T_\nu) = f(x^m; \hat{\theta}_m) \prod_{i=1}^{\nu-1} f(x_{t_i+1}^{t_{i+1}} \mid x^{t_i}; \hat{\theta}_{t_{i+1}}). \tag{9.7}$$

We write $x_i^j = x_i, \ldots, x_j$ and $x^j = x_1^j$. Notice in particular that the ML estimate $\hat{\theta}_{t_{i+1}}$ depends on the entire sequence $x^{t_{i+1}}$, including the last data point $x_{t_{i+1}}$. For instance, for $k = 2$ and $T_2 = \{t_1 = 2, t_2 = 5\}$

$$f(x^5; T_2) = f(x^2; \hat{\theta}_2) f(x_3^5 \mid x^2; \hat{\theta}_5). \tag{9.8}$$

We find the maximizing sequence T_ν and give the result as the sequentially maximized likelihood theorem.

Theorem 9.1 *The sequence that maximizes the density or probability $f(x^n; T_\nu)$, as the case may be, is given by $\{m, m+1, m+2, \ldots, n\}$, and the sequentially maximized likelihood is given by*

$$f(x^n; \hat{\theta}^n) = \max_m f(x^m; \hat{\theta}_m) \prod_{t=m+1}^{n} f(x_t \mid x^{t-1}; \hat{\theta}_t), \tag{9.9}$$

where $\hat{\theta}^n = (\hat{\theta}_1, \ldots, \hat{\theta}_{\hat{m}}, \hat{\theta}_{\hat{m}+1}, \ldots, \hat{\theta}_n)$, and $\hat{m}$ is the maximizing value for m.

Proof For any m consider a factor in (9.7), where we put $\tau = t_i$ and $\tau' = t_{i+1}$:

$$f(x_{\tau+1}, \ldots, x_{\tau'} \mid x^\tau; \hat{\theta}_{\tau'}) = f(x_{\tau'} \mid x^{\tau'-1}; \hat{\theta}_{\tau'}) f(x^{\tau'-1}; \hat{\theta}_{\tau'}). \tag{9.10}$$

The second factor on the right hand side is not greater than

$$f(x^{\tau'-1}; \hat{\theta}_{\tau'-1}) = f(x_{\tau'-1} \mid x^{\tau'-2}; \hat{\theta}_{\tau'-1}) f(x^{\tau'-2}; \hat{\theta}_{\tau'-1}). \tag{9.11}$$

Repeating this until the selected factor is written as the product of one step ahead conditional densities, and doing the same for all the conditional factors in (9.7) we conclude the proof.

Since the batch sequence is a special instance we have shown that the maximized likelihood $f(x^n; \hat{\theta}(x^n))$ is not the maximum! We wish to normalize the conditionals (9.9) to generate recursively a density function $\hat{f}(x^t) = \hat{f}(x^{t-1}) \hat{f}(x_t \mid x^{t-1})$ for $t = \hat{m}+1, \hat{m}+2, \ldots$, where $\hat{m}$ is the value that maximizes (9.6):

$$\hat{f}(x_t \mid x^{t-1}) = \frac{f(x_t \mid x^{t-1}; \hat{\theta}(x^t))}{K(x^{t-1})}, \tag{9.12}$$

$$K(x^{t-1}) = \int f(x_t \mid x^{t-1}; \hat{\theta}(x^t)) dx_t, \tag{9.13}$$

$$\hat{f}(x^{\hat{m}}; \hat{\theta}(x^{\hat{m}})) = f(x^{\hat{m}}; \hat{\theta}(x^{\hat{m}}))/C_{\hat{m}}, \tag{9.14}$$

$$C_{\hat{m}} = \int f(y^{\hat{m}}; \hat{\theta}(y^{\hat{m}})) dy^{\hat{m}}. \tag{9.15}$$

The ranges of the integrals may have to be restricted to be finite. This gives the universal *sequentially normalized maximum likelihood* (SNML) model defining a random process

$$\hat{f}_S(x^n; k) = \hat{f}(x^{\hat{m}}; k) \prod_{t=\hat{m}+1}^n \hat{f}(x_t \mid x^{t-1}) \tag{9.16}$$

$$= \hat{f}(x^{\hat{m}}; k) \prod_{t=\hat{m}+1}^n \frac{f(x_t \mid x^{t-1}; \hat{\theta}(x^t))}{K(x^{t-1})}, \tag{9.17}$$

where we now show the number of parameters. Because the numerator is sequentially maximized, $\hat{f}_S(x^n; k)$ satisfies the necessary condition for optimality at all x^n.

9.1 Bernoulli class

We begin with the Bernoulli class $\mathcal{B} = \{P(x^n; p)\}$, where the parameter $p = P(1)$. The ML estimate is given by $\hat{p}_t = \hat{p}(x^t) = t_1/t$, where t_1 denotes the number of

1s in x^t and $t_0 = t - t_1$. The maximized likelihood is

$$P(x^t; t_1/t) = \left(\frac{t_1}{t}\right)^{t_1} \left(\frac{t_0}{t}\right)^{t_0}. \tag{9.18}$$

Writing $x^t, 0$ for the string of length $t+1$ with the last digit 0, we get the non-normalized conditional probabilities as

$$P(0|x^t; t_1/(t+1)) = \frac{P(x^t, 0; t_1/(t+1))}{P(x^t; t_1/(t+1))} \tag{9.19}$$

$$= \frac{[(t_0+1)/(t+1)]^{t_0+1}[t_1/(t+1)]^{t_1}}{[(t_0+1)/(t+1)]^{t_0}[t_1/(t+1)]^{t_1}} \tag{9.20}$$

$$= (t_0+1)/(t+1), \tag{9.21}$$

where $x^0 = \lambda$ is the empty string. Similarly

$$P(1|x^t; (t_1+1)/(t+1)) = (t_1+1)/(t+1). \tag{9.22}$$

By normalization with

$$K_t = (t_0+1)/(t+1) + (t_1+1)/(t+1) = \frac{t+2}{t+1}, \tag{9.23}$$

we get the conditionals

$$\hat{P}(0|x^t, 0) = \frac{t_0+1}{t+2}, \tag{9.24}$$

$$\hat{P}(1|x^t, 1) = \frac{t_1+1}{t+2}. \tag{9.25}$$

These are seen to equal Laplace's rule with a new interpretation. There are several derivations of Laplace's rule, but none showing that it is the perfectly fair sequential maximum capacity estimator.

The maximum sequential capacity by (9.16) is given by

$$\log \hat{C}_n(S) = \sum_{i=1}^{n+1} \log i!, \tag{9.26}$$

which is a lot bigger than $\log \hat{C}_n = \frac{1}{2}\log(n\pi/2) + o(1)$ in (4.32). However, there are common factors in the numerator and the denominator in (9.16), so that the probabilities of sequentially maximized and ML batch mode strings remain comparable at least for long strings. The SNML probability of the string x^n is given by

$$\hat{P}_S(x^n) = \prod_{t=1}^{n} \hat{P}(x_{t+1} | x^t) = \binom{n}{n_1}^{-1}/(n+1). \tag{9.27}$$

Its negative logarithm is

$$\log 1/\hat{P}_S(x^n) = \log \binom{n}{n_1} + \log(n+1) \tag{9.28}$$

$$= nh(n_1/n) + \frac{1}{2}\log n - \frac{1}{2}\log[2\pi(n_0/n)(n_1/n)] + o(1), \tag{9.29}$$

where $h(p)$ is the binary entropy function, and the second equality holds provided n_1/n is bounded away from 0 and 1.

There is a different way to normalize the likelihood $f(x^n; \hat{\theta}(x^n))$, which was done in [49] for Bernoulli models and discrete Markov chains. In fact, we regard the maximized likelihood $f(x^n; \hat{\theta}(x^n))$ as a function of the last variable x_n only, and normalize the result to give

$$\bar{f}(x_n \mid x^{n-1}) = \frac{f(x^n; \hat{\theta}(x^n))}{\int f(x^n; \hat{\theta}(x^n))dx_n}. \tag{9.30}$$

We get

$$\bar{f}(x_n \mid x^{n-1}) = \frac{f(x^n; \hat{\theta}(x^n))}{\bar{K}(x^{n-1})}. \tag{9.31}$$

The product

$$\bar{f}(x^n) = \hat{f}(x^m; k) \prod_{t=m+1}^{n} \bar{f}(x_t \mid x^{t-1}) \tag{9.32}$$

gives another universal model that defines a random process. By a certain limiting process it gives the conditional probability for binary strings

$$P(x_{t+1} = 0 | x^t) = \frac{t_0 + 1/2}{t+1}, \tag{9.33}$$

usually credited to Krichevski and Trofimov, for which the code length

$$\log 1/P_{KT}(x^n) = \sum_{t} \log 1/P(x_{t+1}|x^t) \tag{9.34}$$

is shorter than that of (9.28) for very skewed strings. For other strings it must exceed (9.28).

9.2 Variable order Markov chains

A most important class of models for discrete alphabets comprises the Markov chains. For simplicity we describe only binary alphabets. To model a binary string $x_1, x_2, \ldots, x_t, \ldots$ define a *state* s_t for the symbol x_{t+1} as the past string

$x_{t-k+1}, \ldots, x_t$ of length k. The state space S consists then of the set of all strings of length k and has 2^k states. At each of them a probability $P(x_{t+1} = 0 \mid s_t), s_t \in S$, is given for the next symbol. The state transition is $s_t, x_{t+1} \mapsto s_{t+1}$, and the probability of the string x^n is given by

$$P(x^n; \theta) = \prod_{t=1}^{n} P(x_t \mid s(x^{t-1}); \theta), \tag{9.35}$$

where $x_t = 0$, for $t < 1$, and $\theta = \{P(0 \mid s), s \in S\}$ denotes the k parameters.

For large alphabets, A the number of parameters that are needed and must be estimated from data grows very fast with the order k, namely as $|A|^k$, and can easily exceed n, which has important consequences. For modeling the English language the alphabet size is, say, 50. If the word length is about 10, and we want to capture the important statistical dependencies in the symbols within words we would need 50^{10} different states, each having 49 probability parameters. It is clear that we cannot afford to take k much larger than 2 or 3. And yet there cannot be more different states in the string than its length even if we take the entire past strings as states. Moreover, the only states that can teach us something about the string's statistical features are the ones that actually occur in the string. This is what led to the *variable order* Markov chains and *tree machines*, introduced in [39], and developed further in [58]. Extensions of these to unbounded trees and states have been studied by A. Galves and his associates [13].

For the binary alphabet a tree machine is defined by a complete binary tree, whose leaves are called *contexts*, each having the probability of the next symbol $P(x_{t+1} = 0 | x_t, x_{t-1}, \ldots, x_{t-k(x^t)})$. These define the parameters θ. The context $c = c(x^t)$ is the path from the root to the leaf, the length depending on the past string. The contexts act as states, although they are not always states. Each context has an extension, which is a state and for which the state transition can be defined just as above. If S is the smallest set of states defined by the set C of leaves, then for $s \in S$ the state transition is defined as $s \mapsto (s, x_t)$, where (s, x_t) is the longest suffix of the concatenate of the string s and the symbol x_t that falls in S. For instance, if $S = 0, 01, 11$, then $(01, 0) = 0$, and $(01, 1) = 11$, and so on. That the contexts are not always states is illustrated in Figure 9.1 by the simplest tree whose leaves are not states (this example is due to M. Weinberger).

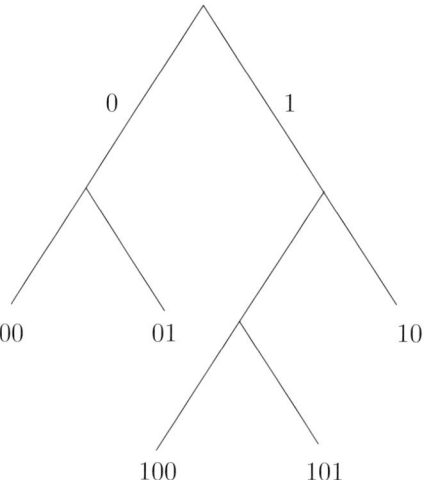

Figure 9.1 The smallest tree without states.

To see this take the string $x^8 = 00101110$. If the initial state is 00, the next symbol 1 takes it into 01. The sequence of states by the longest suffix rule is 00, 01, 010, 11, 11, but for the last symbol 0, there are just three suffixes in the string 110, namely 0, 10, and 110, and none of them is a leaf and hence a state. However, make one more split of the right-most leaf 11 in the tree, which replaces it with two leaves 110 and 111 such that the six leaves are states with a state transition defined on them.

The smallest set of probability parameters θ is still defined at the leaves of a context tree, and the probability of the string x^n is given by

$$P(x^n;\theta) = \prod_{t=|c_0|+1}^{n} P(x_t \mid c(x^{t-1})), \qquad (9.36)$$

where $|c_0|$ denotes the length of the initial context, whose probability is taken as unity.

By the same technique as in the Bernoulli class, the symbol counts at the leaves define normalized conditionals

$$\hat{P}(x_{t+1} = 0 \mid x^t) = \frac{t_{0|c}+1}{2+\sum_i t_{i|c}}, \qquad (9.37)$$

where $t_{i|c}$ denotes the number of times symbol i occurs at context $c = c(t) = c(x^t)$ of C. This differs slightly from the Laplace rule at each context, because $\sum_i t_{i|c}$ differs from t_c by unity when a context c occurs for the first time. The result is

9.2 Variable order Markov chains

a powerful SNML tree machine model for a fixed tree,

$$\hat{P}_S(x^n) = \prod_{t=0}^{n-1} \hat{P}(x_{t+1} \mid x^t), \tag{9.38}$$

where the conditional probabilities at the contexts are given in (9.37).

The generalization to non-binary alphabets is straightforward. We mention to this end that the sequential model $\hat{P}_S(x^n; k)$, with the Laplace predictor at each context, has been applied as an approximation to the NML model [2], for which no closed form solution is known.

9.2.1 Algorithm Context

There is an efficient algorithm which generates both the structure and the probabilities of a finite tree machine from the data. This is done by growing a tree starting with just a one-node tree, the root, which is taken as a leaf.

Algorithm Context

1. Given a string $x_1, x_2, \ldots, x_t x_{t+1}, \ldots$ begin at $t = 0$ with a one-node tree, marked with counts for the two symbols $(n_0, n_1) = (0, 0)$ and code length $L(\lambda) = 0$.
2. Read next symbol $x_{t+1} = i$. If none exists, exit. Otherwise, climb the tree by reading the past string backwards $x_t, x_{t-1}, \ldots$ until the leaf, and update the count n_i for symbol i by unity and the code length $L(s)$ by adding $-\log P(i|s)$, from (9.24), at every node s met until one of the two conditions in step 3 is satisfied.
3. If the node whose count n_i becomes 1 after the update is an internal node, go to 2. But if it is a leaf, create two new nodes and initialize their counts to $n_0 = n_1 = 0$. Go to 2.

In general, this portion of the algorithm creates for a string x^n a lopsided tree, because the tree grows mostly along the paths traveled frequently. Further, each node s represents a "context," in which symbol i occurs about n_i times in the entire string. In fact, since the path from the root to the node s is a binary string $s = i_1, i_2 \ldots, i_k$ the substring $i_k, \ldots, i_1, i$ occurs in the entire string x^n very close to $n(i|s)$ times, where $n(i|s)$ is the count of i at the node s. Notice that the real occurrence count may be a little larger, because the substring may have occurred before node s was created. Also, the following important condition is satisfied

$$n(i|s0) + n(i|s1) = n(i|s), \qquad (9.39)$$

at all nodes whose counts are greater than 0. Finally, the tree created is complete. There exist versions of the algorithm which create incomplete trees.

We give an example of the five trees for the string 00100. Instead of a set of five trees, one after each symbol is read, we write them as five rows, each listing all the nodes in the following order: depth first, then lexical order from left to right: $\lambda; 0, 1; 00, 01; 000, 001$, etc. with semicolons separating depths. Here are the five trees, each followed by the count pairs in the internal nodes, written in the same left-to-right order, while all the leaves have counts (0,0), which are omitted:

1. λ;0,1 count pairs: (1,0);
2. λ;0,1;00,01 count pairs: (2,0);(1,0);
3. λ;0,1;00,01 count pairs: (2,1);(1,0);
4. λ;0,1;00,01,10,11 count pairs: (3,1);(1,0),(1,0);
5. λ;0,1;00,01,10,11;010,011 count pairs: (4,1);(2,0),(1,0);(1,0).

The final tree has four internal nodes λ;0,1;01 with the four count pairs given. Notice that when the symbol 1 is read, the root is an internal node, and the second tree does not grow; only the count of symbol 1 at the root is increased.

9.2.2 Tree pruning

Tree pruning is necessary because the leaf counts of the complete tree after processing the growing string x^t are just (0,0). Since each symbol x_{t+1}, even the first, occurs in at least one node s, we can encode it with the ideal code length given by (9.24) or (9.33), where the counts are those at the node s. If a symbol occurs in node s, other than the root, it also occurs in every shorter node on the path to s, which raises the question of which node we should pick. There are several ways to select this "encoding" node, and we describe two of them. We use the notation $x^t(s)$ for the substring of the symbols of the "past" string x^t that have occurred at the node s and $L(x^t(s))$ for its code length, which is computed using conditional probabilities (9.24) or (9.33). Because the algorithm tends to skew the probabilities at the nodes the latter probability assignment can be better than the Laplace assignment, at least with few counts.

If we are willing to find the encoding nodes in two passes through the data, we can find the optimal subtree of the complete tree, say $\mathcal{T}$, after all the data x^n have been processed. The node counts decrease as the depth of the node increases, and (9.39) insures that when each symbol occurs at a node it also occurs at the son nodes. We write now the code lengths $L(x^t(s))$ more simply as $L(s)$. The pruning is done as follows

1. Initialize: for the tree $\mathcal{T}$ put $\bar{S}$ as the set of the leaves. Calculate $I(s) = L(s)$ at the leaves.
2. Recursively, starting at the leaves compute at each father node s

$$I(s) = \min\{L(s), \sum_{j=0}^{1} I(sj)\}. \tag{9.40}$$

If $L(s)$ is smaller than or equal to the second, replace the sons in the set $\bar{S}$ by the father; otherwise, leave $\bar{S}$ unchanged.

3. Continue until the root is reached.

A second but not necessarily optimal way to select the encoding nodes is to do it in one pass of the data. While growing the tree we can compare the code length at every node passed with the sum of the code lengths of the son nodes, and pick the encoding node for x_{t+1} the first node s such that

$$L(x^t(s)) \leq \sum_{i=0}^{1} L(x^t(si)). \tag{9.41}$$

We mark the so-found node as a special encoding node, which, however, may change when it is passed through the next time. Notice that both sides refer to the code length for the same symbols, the left hand side when the symbols occur at the father node and the right hand side when they occur at the son nodes. The rule then finds the first node where the son nodes' code length is no longer better than the father node's code length. Clearly, because the algorithm does not search the entire path to the leaf such a strategy may not find the very best node. However, since each node s has a unique path from the root to it we can continue comparing its code length with the earlier found encoding node's code length along its path and replace the encoding node's status if needed. This way when the entire string x^n is processed the encoding nodes are found.

Both of the ways to select the encoding nodes after the datum x^n is processed give a complete subtree $\mathcal{T}_n$ with leaves $S = \{s_1, s_2, \ldots, s_K\}$ listed from left to right, each leaf having the counts $(c_0(s), c_1(s))$ with sum $c(s) = c_0(s) + c_1(s)$. Write $\hat{P}_i(s) = c_i(s)/c(s)$. Reorder $S = \{s_{(1)}, s_{(2)}, \ldots, s_{(K)}\}$ by "skewness" $P_0(s_{(i)}) \leq P_0(s_{(i+1)})$. It may be possible to obtain an even smaller tree by partitioning the K leaves in set S into equivalence classes $\Lambda = \{S_1, S_2, \ldots, S_k\}$ such that the ideal code length of x^n is minimized. This can be done by the dynamic programming algorithm due to Bellman in $O(K^2)$ calculations.

In order to simplify the indexing we rewrite the leaf s with the smallest probability $\hat{P}_0(s) = c_0(s)/c(s)$ as s_1 rather than $s_{(1)}$, so that the leaves $S = \{s_1, s_2, \ldots, s_K\}$ are now not listed from left to right but in non-decreasing probabilities $\hat{P}_0(s)$, and in this order the equivalence classes are discrete intervals (τ, t) for $\tau < t$ and $t = 2, 3, \ldots, K$. Let the pooled counts be $C_0(1, t) = \sum_{i=1}^{t} c_0(s_i)$ and $C(1, t) = \sum_{i=1}^{t} (c_0(s_i) + c_1(s_i))$, which define the ideal code length, (9.28),

$$L(\tau, t) = \log \binom{C(\tau, t)}{C_0(\tau, t)} + \log(C(\tau, t+1)). \tag{9.42}$$

The dynamic programming algorithm is defined recursively in $k = 1, 2, \ldots$ by the equations

$$\hat{L}_1(t, K) = L(t, K), \quad t = 1, 2, \ldots, K, \tag{9.43}$$

$$\hat{L}_k(\tau, K) = \min_t [L(\tau, t) + \hat{L}_{k-1}(t+1, K)]. \tag{9.44}$$

Now find the value $\hat{k}$, which minimizes $\min_k \hat{L}_k(1, K)$, and the final partition is defined by the minimizing break points $t_1, t_2, \ldots, t_{\hat{k}}$ in the recursive calculation of $\hat{L}_{\hat{k}}(1, K)$, namely $\hat{L}_1(t_{\hat{k}}, K), \hat{L}_2(t_{\hat{k}-1}, K), \ldots$.

The generalization to non-binary alphabets is similar but more complex, mainly because the pruning of the tree is done somewhat differently so that not all the sons of a node are either retained or removed. An early study of such partitions was done in [54].

9.2.3 Universal code

It is easy to convert Algorithm Context into a universal code. All we need is to have an arithmetic coding unit, which receives as the input the "next" symbol x_{t+1} and the predictive probability (9.24) or (9.33) at its encoding node, say $s^*(t) = s^*(x^t)$, rewritten here

$$P(x_{t+1} = 0|s^*(t)) = \frac{n(0|x^t(s^*(t))) + 1}{|x^t(s^*(t))| + 2}, \qquad (9.45)$$

$$P(x_{t+1} = 0|s^*(t)) = \frac{n(0|x^t(s^*(t))) + 1/2}{|x^t(s^*(t))| + 1}. \qquad (9.46)$$

Notice that only one coding unit is needed – rather than one for every encoding node. This is because an arithmetic code works for any set of conditional probabilities defining a random process, and, clearly, the conditional probabilities of Algorithm Context define a process by

$$P(x^n) = \prod_{t=0}^{n-1} P(x_{t+1}|s^*(t)), \qquad (9.47)$$

where $s^*(0) = \lambda$, the root node.

The algorithm generalizes to non-binary alphabets $\{0, 1, \ldots, d-1\}$ in a straightforward way. The conditional probability rules are generalized thus:

$$P(x_{t+1} = i|x^t) = \frac{t_i + 1}{t + d}, \qquad (9.48)$$

$$P(x_{t+1} = i|x^t) = \frac{t_i + 1/d}{t + 1}. \qquad (9.49)$$

Here is the non-binary Algorithm Context:

1. Begin with a one-node tree, marked with counts $(c_0, \ldots, c_{d-1}) = (0, \ldots, 0)$ and code length $L(\lambda) = 0$.
2. Read the next symbol $x_{t+1} = i$. If none exists, exit. Otherwise, climb the tree by reading the past string backwards $x_t, x_{t-1}, \ldots$ and update the count n_i by unity and the code length $L(s)$ by $-\log P(i|s)$ obtained from (9.48) at every node s met until one of the two following conditions is satisfied:
3. If the node whose count n_i becomes 1 after the update is an internal node, go to 2. But if it is a leaf, create d new nodes and initialize their counts to $n_i = 0$. Go to 2.

This time when the counts are positive the father node's count of each symbol exceeds the sum of the sons' counts by unity:

$$\sum_j n(i|sj) + 1 = n(i|s). \tag{9.50}$$

The selection of the encoding node may be done in the same way as in the binary case, but there exist other more sophisticated strategies, which are needed to find useful partitions. We do not discuss these here.

What is the code length of this universal code? To answer this question we assume that the data are generated by a tree machine with K leaves. One can show [58] that the mean ideal code length resulting from the universal code, defined by a modification of the rule (9.40), satisfies

$$\frac{1}{n} EL_{TM}(X^n) = H(X) + \frac{K}{2n} \log n + o(n^{-1} \log n),$$

where $H(X)$ is the entropy of the tree model generating the data. Moreover, by the consistency theorem in Section 5.2 no universal code exists where the mean per symbol code length approaches the entropy faster.

9.2.4 Extension to time series

Since the tree grows only at repeated symbol occurrences the algorithm given will not produce other than depth one trees for sequences over a large alphabet such as resulting from quantized real numbers unless the sequences are enormously long. The solution to this problem is to use two quantizations, one a coarse precision for the tree growth and the other for encoding the high precision data at the nodes. We may consider the leaves as providing a partition of the space of sequences, each node defining its own predictor or encoder depending on the application.

Consider a high precision data string $x^n = x_1, x_2, \ldots, x_n$. Let $x \mapsto \bar{x}$ denote the map when x is quantized to a low precision number, perhaps even binary: $\bar{x} = 1$, if $x > 0$, and $\bar{x} = 0$, if $x \leq 0$. We apply the tree growth algorithm above to the data $\bar{x}^n = \bar{x}_1, \bar{x}_2, \ldots, \bar{x}_n$.

Before considering the pruning of the tree we assign a probability function to the nodes s, for instance, a gaussian density function quantized to the precision in which the original data were given. Hence, we have a conditional probability function $P(x_{t+1}|x^t; \hat{x}^{t+1}, \hat{\tau}_s)$, where $\hat{\tau}_s$ denotes the estimate of the variance obtained from the data points x_{t+1} when $\bar{x}_t \bar{x}_{t-1} \ldots = s$, and, for instance, for well-known AR models we may write

$$\hat{x}^{t+1} = \hat{a}_{s0} + \hat{a}_{s1}x_t + \hat{a}_{s2}x_{s(t-1)} + \cdots + \hat{a}_{sk}x_{s(t-k-1)}, \tag{9.51}$$

where the coefficients are determined from the occurrence counts of data at s by the least squares technique.

Since each node selects data of a certain narrow type, it may well happen that the covariance matrix needed to get the coefficients $\hat{a}_{si}$ by minimization is ill conditioned, and a suitable technique such as singular component decomposition is needed to find the minimizing parameters. An interesting phenomenon occurs if the order k is just one symbol larger than the path length to the node s. The effect is like smoothing a piecewise linear partition of the context space.

The pruning of the tree may be done just as described above, or it may be done in terms of the minimized quadratic criterion by which the coefficients $\hat{a}_{si}$ are determined. The result is a universal non-linear AR time series model.

Example This work was done with Gilbert Furlan. The data $x^n = x_1, \ldots, x_{500}$ were generated by the non-linear equation

$$x_{t+1} = -.22|x_t|^{1/2}x_{t-1} + .1|x_{t-2}|^{1.3} + e_{t+1}, \tag{9.52}$$

where e_t was iid gaussian noise of zero mean and variance $\tau_e = 4.25$. This gave the variance $\tau_x = 6.18$ to the x-process.

We constructed the universal context model above with the coarse precision as binary, and with a tree depth varying from 2 to 3 we predicted the data with the linear least squares predictor (9.51), with the coefficients determined from all the data at each leaf. The prediction error turned out to be $\hat{\tau}_x = 5.30$. We also generated a new batch of data y^n of the same length, and with the tree and the coefficients determined from the original data we calculated the prediction error. It came out as $\hat{\tau}_y = 4.85$. Clearly, the new data happened to be easier to predict. In fact, the original system is close to instability, and the sample data vary quite a bit. Finally, we fit an AR model by the predictive principle, which gave $\hat{k} = 2$ as the optimal order, and the prediction error for the original data as 5.72. We then applied the optimal linear AR model to the new data y^n and got

5.39 as the prediction error. The universal tree model clearly performed better than the optimal linear model.

Algorithm Context applies to chaotic processes and easily solves most of the problems of interest including the estimation of the parameters [43]. Here is a simple example.

Example The so-called logistic map defined by the recursion

$$x_{t+1} = 4x_t(1 - x_{t-1}) + e_{t+1} \tag{9.53}$$

is an example of a chaotic process, when the data x_t are quantized to any precision, such that e_t may be viewed as the quantization error. The function $4x(1-x)$ is a concave quadratic function with the maximum 1 at $x = 1/2$, which vanishes at $x = 0$ and $x = 1$.

We applied Algorithm Context to the binary data $\bar{x}^n$, where $\bar{x}_i = 0$, when $x_i \in (0, 1/2]$, and $\bar{x}_i = 1$, when $x_i \in (1/2, 1]$, and grew the tree only to the balanced tree of depth 2. Actually, the tree could have been optimized, but this was not done. The four leaves partition the data into four equivalence classes by $\{x^t : x_{t-1}, x_t \mapsto \bar{x}_{t-1}, \bar{x}_t\}$ or, in words, all sequences for which the last two symbols are mapped to one of the four corresponding nodes $s = \bar{x}_{t-1}, \bar{x}_t$ are equivalent. To the data $\{x_{st}\}$ at each node or in the corresponding equivalence class fit an AR model of optimal order by

$$\min_{\{a_i\}} \sum_i (x_{st} - a_0 - a_{s1}x_{s(t-1)} - \ldots)^2. \tag{9.54}$$

The result is a non-linear representation of the data:

$$x_{t+1} = \hat{a}_{s0} + \hat{a}_{s1}x_{st} + \hat{a}_{s2}x_{s(t-1)} + \hat{a}_{s3}x_{s(t-2)} + \epsilon_{t+1} = \hat{F}(x_t) + u_{t+1}, \tag{9.55}$$

where the index s refers to the node determined by $\bar{x}^t$, and the function $\hat{F}$ is defined by expressing x_{t-1} and x_{t-2} in terms of x_t and putting the terms x_{t-i} for $i > 2$ within u_{t+1}. Quite remarkably, the function $\hat{F}(x)$ is a virtual duplicate of $4x(1-x)$; it is slightly narrower. One would perhaps expect a piecewise linear approximation of the quadratic function.

9.3 Linear quadratic regression models

In this section, which is an extract from [46] without the harder asymptotics, we consider the linear-quadratic regression problem for the data y^n, X_n modeled as follows:

$$y_t = b'\bar{x}_t + \epsilon_t, \tag{9.56}$$

where the prime indicates transposition, $b' = (b(1), \ldots, b(k))$, and $\bar{x}_t$ are the columns of the regressor matrix X_n consisting either of fixed numbers, not defined by y^n, or $\bar{x}_t = col\{y_{t-1}, \ldots, y_{t-k}\}$ for $t > k$ as in AR models. The deviations $\{\epsilon_t\}$ are taken as an iid sequence generated by a normal distribution of zero mean and variance σ^2. These induce the normal density function for the data $f(y^n \mid X_n; \sigma^2, b) = (2\pi\sigma^2)^{-n/2} e^{-\frac{1}{2\sigma^2} \sum_1^n (y_t - b'\bar{x}_t)^2}$.

Consider the three representations of data

$$y_t = b'_n \bar{x}_t + \hat{\epsilon}_t(n) = \sum_{i=1}^k b_{n,i} x_{t,i} + \hat{\epsilon}_t(n), \; t \geq m, \tag{9.57}$$

$$y_t = b'_{t-1} \bar{x}_t + e_t = \sum_{i=1}^k b_{t-1,i} x_{t,i} + e_t, \; t > m, \tag{9.58}$$

$$y_t = b'_t \bar{x}_t + \hat{e}_t = \sum_{i=1}^k b_{t,i} x_{t,i} + \hat{e}_t, \; t \geq m, \tag{9.59}$$

where m is the smallest number for which the ML estimates b_t are defined. These are given by the well-known recursions, see, for instance, [33],

$$b_t = \hat{\theta}(y^t, X_t) = V_t \sum_{j=1}^t \bar{x}_j y_j \tag{9.60}$$

$$= b_{t-1} + V_{t-1} \bar{x}_t (y_t - \bar{x}'_t b_{t-1}) / (1 + c_t), \tag{9.61}$$

$$V_t = (X_t X'_t)^{-1} = V_{t-1} - V_{t-1} \bar{x}_t \bar{x}'_t V_{t-1} / (1 + c_t), \tag{9.62}$$

$$c_t = \bar{x}'_t V_{t-1} \bar{x}_t, \tag{9.63}$$

$$d_t = \bar{x}'_t V_t \bar{x}_t, \tag{9.64}$$

$$1 - d_t = 1/(1 + c_t). \tag{9.65}$$

The last equality was shown in [24] and [57] with the interpretation that the quantity $1 - d_t$ is the ratio of the (Fisher) information in the first $t - 1$ observations relative to all the t observations [57].

The predictor $\bar{x}'_t b_{t-1}$ of y_t is sometimes called the "plug-in" predictor, in which the parameters b are replaced by the ML estimates, calculated from the data available up to $t-1$. The resulting model (9.58) is widely studied, [9], [24], [42], [57], and it defines the *predictive* linear least squares (PLS) model and the more general predictive minimum description length (PMDL) model. The third representation (9.59) is the one we are interested in, and the first (9.57) is traditional.

The main objective in these regression problems is to get statistical information from the data by minimizing the quadratic loss function. However, in the present theory we represent the loss function in terms of the density the models assign to the data.

As in Theorem 9.1 let $T_\nu = \{t(1) = m < t(2) < \cdots < t(\nu) = n\}$, where it now will be more convenient to write $t_i = t(i)$, and consider the sum of the squared residuals

$$\hat{s}_n(T_\nu) = s_m + \sum_{i=1}^{\nu-1} \sum_{j=t(i)+1}^{t(i+1)} (y_j - \bar{x}'_j b_{t(i+1)})^2, \qquad (9.66)$$

where

$$s_t = \sum_{1}^{t}(y_j - b'_t \bar{x}_j)^2. \qquad (9.67)$$

The minimizing set will be seen to be $T_\nu = \{m, m+1, \ldots, n\}$, and with the notation (9.59) the minimized sum of the squares for $m \leq t \leq n$, all $n \geq m$, is given by

$$\hat{s}_t = s_m + \sum_{j=m+1}^{t}(y_j - \bar{x}'_j b_j)^2 = s_m + \sum_{j=m+1}^{t} \hat{e}_j^2 = \hat{s}_{t-1} + \hat{e}_t^2, \qquad (9.68)$$

the last equation holding for $t > m$. Assuming for the sake of notational simplicity that $t(2) - t(1) > 1$ we see the minimality of $\hat{s}_t$ by calculating recursively for $t = m+1, m+2, \ldots, t(2)$

$$\hat{s}_{m+1} = s_m + \hat{e}_{m+1}^2 = s_{m+1}, \qquad (9.69)$$

$$\hat{s}_{m+2} = \hat{s}_{m+1} + \hat{e}_{m+2}^2 < s_{m+2},$$

$$\hat{s}_{m+3} = \hat{s}_{m+2} + \hat{e}_{m+3}^2 < s_{m+3},$$

$$\vdots$$

$$\hat{s}_{t(2)} < s_{t(2)}. \qquad (9.70)$$

The inequalities hold, because $s_t < s_m + \sum_{m+1}^{t}(y_j - \bar{x}'_j b_{t+1})^2$.

9.3.1 Fixed variance

Consider first the case where the variance σ^2 is fixed. The non-normalized conditionals are given by

$$f(y_t \mid y^{t-1}, X_t; \sigma^2, b_t) = f(y_t \mid X_t; \sigma^2, b_t) = \frac{1}{\sqrt{2\pi\sigma^2}} \exp\left(-\frac{(y_t - \hat{y}_t)^2}{2\sigma^2}\right), \quad (9.71)$$

where

$$\hat{y}_t = \bar{x}_t' b_t. \quad (9.72)$$

The normalized conditional is

$$\hat{f}(y_t \mid y^{t-1}, X_t; \sigma^2) = \frac{f(y_t \mid X_t; \sigma^2, b_t)}{K(y^{t-1}; \sigma^2)}, \quad (9.73)$$

where

$$K(y^{t-1}; \sigma^2) = \frac{1}{\sqrt{2\pi\sigma^2}} \int_{-\infty}^{\infty} \exp\left(-\frac{(y_t - \hat{y}_t)^2}{2\sigma^2}\right) dy_t. \quad (9.74)$$

By (9.60)

$$\hat{y}_t = \bar{x}_t' \left[V_{t-1}\bar{x}_t(y_t - \bar{x}_t' b_{t-1})/(1 + c_t) + b_{t-1}\right] \quad (9.75)$$

and

$$y_t - \hat{y}_t = \hat{e}_t = \frac{1}{1 + c_t} e_t = (1 - d_t) e_t, \quad (9.76)$$

which is seen to be a multiple of the plug-in prediction error. With (9.76) the integral becomes

$$\int_{-\infty}^{\infty} \exp\left(-\frac{(y_t - \bar{x}_t' b_{t-1})^2}{2((1 + c_t)\sigma)^2}\right) dy_t = \sqrt{2\pi(1 + c_t)^2 \sigma^2}. \quad (9.77)$$

Finally, the normalized conditional density is the Gaussian density with the mean given by the plug-in predictor, and the variance $\tau^2 = (1 + c_t)^2 \sigma^2$:

$$f(y_t \mid y^{t-1}, X_t; \sigma^2, b_t) = \frac{1}{\sqrt{2\pi\tau^2}} \exp\left(-\frac{(y_t - \bar{x}_t' b_{t-1})^2}{2\tau^2}\right). \quad (9.78)$$

9.3.2 Free variance

In the model with σ^2 as a free parameter the normalization required in the SNML model is not straightforward, but another closely related construction admits a very simple normalization without any hyperparameters. Consider the maximization problem

$$\max_{\sigma^2} \prod_{t=m+1}^{n} f(y_t \mid y^{t-1}, X_t; \sigma^2, b_t), \quad (9.79)$$

where
$$f(y_t \mid y^{t-1}, X_t; \sigma^2, b) = \frac{f(y^t \mid X_t; \sigma^2, b)}{f(y^{t-1} \mid X_{t-1}; \sigma^2, b)}. \tag{9.80}$$

With the solution $\hat{\sigma}_n^2 = (1/n)\hat{s}_n$ the non-normalized and normalized conditional density functions for $t = m+1, m+2, \ldots$ become

$$f(y_t \mid y^{t-1}, X_t) = \frac{\hat{s}_t^{-t/2}}{\hat{s}_{t-1}^{-(t-1)/2}} = \hat{s}_{t-1}^{-1/2}\left(1 + \frac{(y_t - \hat{y}_t)^2}{\hat{s}_{t-1}}\right)^{-t/2} \tag{9.81}$$

$$\hat{f}(y_t \mid y^{t-1}, X_t) = \hat{s}_{t-1}^{-1/2}\frac{\left(1 + \frac{(y_t - \hat{y}_t)^2}{\hat{s}_{t-1}}\right)^{-t/2}}{K(y^{t-1})} \tag{9.82}$$

$$K(y^{t-1}) = \hat{s}_{t-1}^{-1/2}\int_{-\infty}^{\infty}\left(1 + \frac{(y_t - \hat{y}_t)^2}{\hat{s}_{t-1}}\right)^{-t/2} dy_t. \tag{9.83}$$

To get the normalizing integral write first

$$\hat{y}_t = \bar{x}_t' b_t = d_t y_t + \bar{y}_t, \tag{9.84}$$

$$d_t = \bar{x}_t' V_t \bar{x}_t, \tag{9.85}$$

$$\bar{y}_t = \bar{x}_t' V_t \sum_1^{t-1} y_i \bar{x}_i, \tag{9.86}$$

where $\bar{y}_t$ does not depend on y_t. Then

$$K(y^{t-1}) = \hat{s}_{t-1}^{-1/2}\int_{-\infty}^{\infty}\left[1 + \frac{(1-d_t)^2}{\hat{s}_{t-1}}\left(y - \frac{\bar{y}_t}{1-d_t}\right)^2\right]^{-t/2} dy. \tag{9.87}$$

By change of variables

$$z = \frac{1-d_t}{\sqrt{\hat{s}_{t-1}}}\left(y - \frac{\bar{y}_t}{1-d_t}\right), \tag{9.88}$$

we get for $K_{t-1} = K(y^{t-1})$

$$K_{t-1} = \frac{1}{1-d_t}\int_{-\infty}^{\infty}(1+z^2)^{-t/2} dz = \frac{\sqrt{\pi}}{1-d_t}\Gamma\left(\frac{t-1}{2}\right)/\Gamma(t/2), \tag{9.89}$$

where the last equality comes from the fact that z is seen to have Student's distribution.

The conditional density function is then given by

$$\hat{f}(y_t \mid y^{t-1}, X_t) = \frac{\hat{s}_{t-1}^{-1/2}}{K_{t-1}}\left(1 + \frac{(1-d_t)^2}{\hat{s}_{t-1}}\left(y_t - \frac{\bar{y}_t}{1-d_t}\right)^2\right)^{-t/2} \tag{9.90}$$

$$= K_{t-1}^{-1}\frac{\hat{s}_t^{-t/2}}{\hat{s}_{t-1}^{-(t-1)/2}}. \tag{9.91}$$

Again the predictor that maximizes the conditional density function is seen to be the plug-in predictor, and neither the observed value y_t nor the last component of b_t improves prediction.

To get the density function for all the data y^n the initial density function $q(y^m \mid X_m)$ is needed. We should take it as the NML model $\hat{f}(y^m \mid X_m; \mu_1, \mu_2)$, where μ_1 and μ_2 are certain hyperparameters which can be optimized as explained in Chapter 8, but with which we are not concerned here, and we just write $q(y^m \mid X_m) = \hat{f}(y^m \mid X_m)$ giving $\log 1/\hat{f}(y^m \mid X_m) = (m/2)\log \hat{s}_m + \log C_{m,k}$. We then get the SNLS model

$$\hat{f}(y^n \mid X_n) = q(y^m \mid X_m) \prod_{t=m+1}^{n} \hat{f}(y_t \mid y^{t-1}, X_t). \tag{9.92}$$

By ignoring the initial density and applying Stirling's formula to the gamma functions we get

$$\ln 1/\hat{f}(y^n \mid X_n) = \frac{n}{2}\ln(2\pi e \hat{\tau}_n) \tag{9.93}$$

$$+ \sum_{m+1}^{n} \ln(1 + c_t) + \frac{1}{2}\ln n + O(1), \tag{9.94}$$

where $\hat{\tau}_n = (\hat{s}_n - \hat{s}_m)/(n - m)$.

Theorem 9.2 *If the regressor variables are non-random satisfying (9.111) and the data generated by (9.56), then*

$$\frac{1}{n-m}\sum_{t=m+1}^{n} E\hat{e}_t^2 = \sigma^2\left(1 - \frac{1}{n-m}\sum_{t=m+1}^{n} d_t\right), \tag{9.95}$$

$$\frac{1}{n-m}\sum_{t=m+1}^{n} Ee_t^2 = \sigma^2\left(1 + \frac{1}{n-m}\sum_{t=m+1}^{n} c_t\right), \tag{9.96}$$

$$\frac{1}{n-m}\sum_{t=m+1}^{n} E\hat{\epsilon}_t^2(n) = \sigma^2, \tag{9.97}$$

where the expectation is with the parameters b and σ.

Proof To obtain (9.96) we start with $y_i = \bar{x}_i' b + \epsilon_i$:

$$e_t = y_t - \bar{x}_t' V_{t-1} \sum_{t=1}^{t-1} \bar{x}_i y_i \tag{9.98}$$

$$= \epsilon_t - \bar{x}_t' V_{t-1} \sum_{t=1}^{t-1} \bar{x}_i \epsilon_i, \tag{9.99}$$

and noticing that $\bar{x}_t' V_{t-1} \sum_{i=1}^{t-1} \bar{x}_i \bar{x}_i' b = \bar{x}_t' b$. Further by the fact that $\{\epsilon_t\}$ is a zero-mean iid sequence

$$Ee_t^2 = \sigma^2 (1 + \bar{x}_t' V_{t-1} \sum_{i=1}^{t-1} \bar{x}_i \bar{x}_i' V_{t-1} \bar{x}_t) = \sigma^2 (1 + c_t). \tag{9.100}$$

To derive (9.95) we have as above

$$\hat{e}_t = \epsilon_t - \bar{x}_t' V_t \sum_{t=1}^{t} \bar{x}_i \epsilon_i, \tag{9.101}$$

$$E\hat{e}_t^2 = \sigma^2 [(1 - d_t)^2 + \bar{x}_t' V_t \sum_{i=1}^{t-1} \bar{x}_i \bar{x}_i' V_t \bar{x}_t] \tag{9.102}$$

$$= \sigma^2 [1 + d_t^2 - 2d_t + d_t - d_t^2] \tag{9.103}$$

$$= \sigma^2 (1 - d_t), \tag{9.104}$$

where we first added the term $d_t^2 = \bar{x}_t' V_t \bar{x}_t \bar{x}_t' V_t \bar{x}_t$ in the brackets, which made the sum term d_t, and then we subtracted the same term d_t^2.

To prove the remaining equation, (9.97), we have as in the previous case

$$\hat{\epsilon}_t(n) = \epsilon_t - \bar{x}_t' V_n \sum_{i=1}^{n} \bar{x}_i \epsilon_i \tag{9.105}$$

$$= (1 - \mu_t) \epsilon_t - \bar{x}_t' V_n \sum_{i \neq t} \bar{x}_i \epsilon_i, \tag{9.106}$$

$$E\hat{\epsilon}_t^2(n) = \sigma^2 [(1 - \mu_t)^2 + \bar{x}_t' V_n \sum_{i \neq t} \bar{x}_i \bar{x}_i' V_n \bar{x}_t] \tag{9.107}$$

$$= \sigma^2 (1 - \mu_t), \tag{9.108}$$

where $\mu_t = \bar{x}_t' V_n \bar{x}_t = tr(V_n \bar{x}_t \bar{x}_t')$. Here we have also added and subtracted the term μ_t^2 in the brackets. Then

$$\sum_{t=1}^{n} E\hat{\epsilon}_t^2(n) = \sigma^2 (n - k). \tag{9.109}$$

In the same way

$$\sum_{t=1}^{m} E\hat{\epsilon}_t^2(m) = \sigma^2 (m - k), \tag{9.110}$$

and (9.97) follows concluding the proof.

We conclude with the large data set behavior of the SNLS model. Provided the regressor variables $\bar{x}_t$ satisfy

$$\frac{1}{n}\sum_{1}^{n}\bar{x}_i\bar{x}_i' \to \Sigma, \qquad (9.111)$$

then

$$\ln 1/\hat{f}(y^n \mid X_n) = \frac{n}{2}\ln(2\pi e\hat{\tau}_n) + (1+o(1))\left(k+\frac{1}{2}\right)\ln n. \qquad (9.112)$$

To obtain this we get from (9.111) first $(1/t)V_t \to \Sigma$, and then $\ln(1+c_t) = c_t + O(1/t^2)$. By the first of the following results, derived in [42] and [57],

$$\frac{1}{\ln n}\sum_{t=1}^{n} c_t \to k, \qquad (9.113)$$

$$\frac{1}{\ln n}\sum_{t=1}^{n} d_t \to k, \qquad (9.114)$$

we deduce (9.112). The important step in the derivation is the fact that by (9.63) $c_t = \bar{x}_t' V_{t-1} \bar{x}_t = tr[\bar{x}_i \bar{x}_i' V_{t-1}]$, which by summing implies the claim.

The proofs of consistency in the almost sure sense for general sequential models are quite difficult even for AR models; see [57] for the predictive least squares criterion. Using the results in this reference we obtain the following by martingale arguments [46]:

$$\ln 1/\hat{f}(y^n \mid X_n) = \frac{n}{2}\ln(2\pi e\sigma^2) + (1+o(1))\left(k+\frac{1}{2}\right)\ln n, \qquad (9.115)$$

almost surely for the SNLS AR model, when the data are generated by an AR model.

9.4 Autoregressive moving average (ARMA) models

A most important class of linear regression models are the ARMA models arising when each time indexed random variable x_t is expressed in the form

$$x_t + \sum_{i=1}^{p} a_i x_{t-i} = e_t + \sum_{i=1}^{q} b_i e_{t-i}, \qquad (9.116)$$

where e_t is a normal, independent sequence with mean zero and variance σ^2. Here, the uncorrelated error process in AR models is replaced by the correlated process $e_t + b_1 e_{t-1} + \cdots + b_q e_{t-q}$.

We consider two problems. The first is prediction, i.e. the computation of the optimal predictor in a convenient recursive way, which is solved in [20] for the case where the parameters are known. When the parameters are not known they must be estimated. This requires the predictor equations, which are in an irreducible form that has a minimum number of parameters. The usual predictors have a huge number of excess parameters, and are inappropriate for parameter estimation. The ARMA models are much better, and, moreover, their optimal recursive predictor equations are simpler.

The prediction error for the recursive predictors is too optimistic when the usual assumption of stationarity with large quantities of available data is not good. We conclude the section with a corollary of Theorem 5.3 giving a lower bound for the prediction error for data, generated by any ARMA model in a family for any number of data. An example would be to partition the data sequence into a suitable number of segments and fit the parameters to each segment separately, and then construct predictors.

9.4.1 Prediction

In this section we discuss briefly but in a self-contained way the recursive least squares prediction theory for ARMA processes. Our development of the theory, which is seen to generalize the work on stationary predictors by Whittle, differs from the general Kalman prediction theory in that the Riccati equation with its generally redundant variables is replaced by a covariance factorization algorithm. The resulting predictors for the ARMA processes are not only simpler but they also involve only parameters which can be estimated from the observations. This, of course, is mandatory for the subsequently discussed prediction problem where all the parameters must be estimated. We begin the study of prediction from scratch without assuming recursiveness as in ARMA processes, which we assume later.

Consider a random vector $x' = (x_0, \ldots, x_n)$, with zero mean and the covariance matrix $R = E(xx') = \{r(i,j)\}$, where as usual the "prime" denotes the transposition. We have: (i) $R \geq 0$ (positive semidefinite) always, and (ii) $R > 0$ (positive definite) if and only if the components in x are linearly independent. The proof is easy. (i) Consider the real numbers a_i, $i = 0, \ldots, n$. Then $0 \leq E(\sum a_i x_i)^2$.

(ii) The random variables are linearly independent if and only if $\sum a_i x_i = 0$ implies $a_i = 0$, for all i, or, equivalently, $E(\sum a_i x_i)^2 = 0$ implies $a_i = 0$, for all i.

Let $R > 0$. We orthogonalize the sequence x with the Gram–Schmidt procedure as follows:

$$x_0 = b(0,0)e_0, \tag{9.117}$$

$$x_1 = b(1,1)e_1 + b(0,1)e_0, \tag{9.118}$$

$$\vdots$$

$$x_n = b(n,n)e_n + \ldots + b(0,n)e_0, \tag{9.119}$$

where $e = e_0, \ldots, e_n$ is orthonormal. Equivalently, this can be done by a factorization of the covariance matrix as follows

$$R = B'B,$$

where B is upper triangular,

$$b = \begin{pmatrix} b(0,0) & b(0,1) & \ldots & b(0,n) \\ 0 & b(1,1) & \ldots & b(1,n) \\ \vdots & \vdots & \ddots & \vdots \\ 0 & 0 & \ldots & b(n,n) \end{pmatrix}. \tag{9.120}$$

This is known as the Cholesky factorization. By comparison of the elements we get the recursion

$$b(0,0) = +\sqrt{r(0,0)}, \; b(0,1) = r(0,1)/b(0,0), \tag{9.121}$$

$$b(1,1) = +[r(1,1) - b(0,1)^2]^{1/2}, \tag{9.122}$$

$$\vdots$$

The algorithm works because $R > 0$, which is equivalent with all the principal minors being positive: $|R_t| = b(0,0)^2 \times \cdots \times b(t,t)^2 > 0$, where R_t is the submatrix of R defined by the first t rows and columns.

Before addressing the ARMA prediction problem, we discuss the special case of predicting stationary processes from the infinite past, which was the case studied by Whittle and which back in the mid-1960s gave us the idea of how to solve the issue of recursion. Consider a stationary random process $\{x_t : t = \ldots, -1, 0, 1, \ldots\}$ with zero mean. The set of all finite linear combinations of the random variables in the process is a linear manifold. With $(x, y) = E(xy)$ as the inner product, and $\sqrt{(x,x)}$ as the induced norm, the process defines a Hilbert space as the closure of the linear manifold. We denote by X_t the subspace spanned by $\{x_i | i \leq t\}$. Let $x_{t|t-1}$ denote the orthogonal projection of x_t onto X_{t-1}. Then by successive projections,

$$x_t = e_t + x_{t|t-1} = e_t + p_1 e_{t-1} + \cdots + p_n e_{t-n} + x_{t|t-n-1}, \qquad (9.123)$$

where e_i is an uncorrelated sequence with some variance σ^2. If $x_{t|t-n} \to 0$, as $n \to \infty$, x_t is expanded into the decomposition

$$x_t = e_t + p_1 e_{t-1} + \cdots + p_n e_{t-n} + \cdots, \qquad (9.124)$$

which is also known as the *one-sided moving average* representation. The sequence e_t is called the *innovation* sequence. The important thing here is that the e-sequence is uncorrelated and that the variables e_i, $i \leq t$, also span X_t. We emphasize the second condition, which is needed.

Suppose next that the stationary process is generated by the recursion

$$x_t + a_1 x_{t-1} + \cdots + a_p x_{t-p} = e_t + b_1 e_{t-1} + \cdots + b_q e_{t-q}, \qquad (9.125)$$

for $t = \ldots, -1, 0, 1, 2, \ldots$, where e is an orthogonal zero-mean process with variance $E(e_t^2) = \sigma^2$. The process clearly is stationary if and only if the roots of the polynomial $a(z) = 1 + a_1 z + \cdots + a_p z^p$ are strictly outside the unit circle. Further, as readily seen, x_t admits a one-sided moving average representation with e_t as its innovation sequence if the roots of the polynomial $b(z) = 1 + b_1 z + \cdots + b_q z^q$, too, are outside the unit circle. In that case the linear spaces X_t and E_t, spanned by $\{x_i | i \leq t\}$, and $\{e_i | i \leq t\}$, respectively, are the same. Such a process is called a stationary ARMA(p,q) process.

Let $x_{t|t-1}$ denote the orthogonal projection of x_t on X_{t-1}. Then by the basic projection theorem in Hilbert spaces

$$E[x_t - x_{t|t-1}]^2 \leq E[x_t - \sum_{i=1}^{\infty} c_i e_{t-i}]^2 \qquad (9.126)$$

for all square summable coefficients c_i. The equality is attained at $c_i = p_i$,

$$x_{t|t-1} = \sum_{i=1}^{\infty} p_i e_{t-i}, \qquad (9.127)$$

where p_i is defined by the impulse response of the system (9.125). This projection is called the linear least squares predictor of x_t, because it minimizes the variance of the prediction error

$$e_t = x_t - x_{t|t-1}. \tag{9.128}$$

The prediction problem of practical importance is to express $x_{t|t-1}$ in terms of the past observations $x_{t-1}, x_{t-2}, \ldots$, rather than the innovations, in an as simple to calculate form as possible, which turns out to be easy enough.

Consider first the simple case

$$x_t + a x_{t-1} = e_t + b e_{t-1}, \tag{9.129}$$

where $|a| < 1$ and $|b| \leq 1$, as required for stationarity and for e to be the innovation sequence. In order to insure stability of the predictor we actually require $|b| < 1$. Notice that even if $|b| > 1$, x_t still admits a one-sided moving average representation, but with another sequence as the innovations process. As an exercise the reader may wish to find the innovation sequence when $a = 1/2$ and $b = 2$. Because $X_t = E_t$,

$$x_{t|t-1} = -a x_{t-1} + b e_{t-1}. \tag{9.130}$$

Hence, $e_t = x_t + a x_{t-1} - b e_{t-1}$, and using (9.128) to eliminate e_{t-1} we get

$$x_{t|t-1} + b x_{t-1|t-2} = (b - a) x_{t-1}, \tag{9.131}$$

which solves our problem.

The same arguments give in general

$$\begin{aligned} x_{t|t-1} + b_1 x_{t-1|t-2} + \cdots + b_q x_{t-q|t-q-1} \\ = (b_1 - a_1) x_{t-1} + \cdots + (b_k - a_k) x_{t-k}, \end{aligned} \tag{9.132}$$

where k is the maximum of p and q. Also $b_i = 0$ if $i > q$ and $a_i = 0$ if $i > p$.

The same general idea works even in the more difficult case, where we want to compute recursively the orthogonal projections on the finite dimensional linear space $X_{0,t}$, spanned by the observations $x_0, \ldots, x_t$, and $X_t \neq E_t$.

There exist very fast algorithms for doing this, the so-called fast predictor algorithms, developed in the early 1970s, [34], [35]; see also [19] and [27] for further developments. However, our goal is only to derive the "ordinary" recursive algorithm as a replacement for the recursions involving the Riccati equation in

Kalman's theory. Consider a process, defined for $t \geq p$ by the recursion

$$x_t + a_1 x_{t-1} + \cdots + a_p x_{t-p} = u_t, \tag{9.133}$$

where u_t is a process with zero mean, whose covariance function $E(u_t u_s) = r(t,s)$ satisfies the crucial "bandedness" property

$$r(t,s) = 0 \text{ for } |t-s| > q. \tag{9.134}$$

In addition, we let the initial variables be correlated as follows:

$$E(u_t x_s) = \sigma(t,s) = 0 \quad \text{if } t-s > q, \tag{9.135}$$

$$E(x_t x_s) = \sigma(t,s) \quad t, s \leq p. \tag{9.136}$$

A typical example is the ARMA process resulting when u_t is defined as the moving average (MA) process

$$u_t = e_t + b_1 e_{t-1} + \cdots + b_q e_{t-q}, \tag{9.137}$$

where e_t is an uncorrelated zero-mean process. But the properties (9.136) hold even when a so-called Gauss–Markov process (a state space representation of a linear stochastic system) is converted to its input–output representation by elimination of the state process, in which case the input process is a linear combination of two uncorrelated noise processes. Hence, to cover all the important cases we must not assume that the input process is an MA process (9.137), but rather one which satisfies the weaker conditions (9.136).

The problem now is to find the orthogonal projection of x_t on $X_{0,t-1}$. We still write the projection as $x_{t|t-1}$. What makes this problem different from the previous case is the fact that plainly $X_{0,t-1} \neq U_{p,t-1}$, the latter space being spanned by the input sequence from p up to $t-1$. The obvious remedy is to augment the input sequence with the first p observations, which makes the result span $X_{0,t-1}$. All we then need is to orthogonalize the augmented sequence $v_0, \ldots, v_{t-1}$ with the Gram–Schmidt algorithm, and we have reduced the problem to the previously treated case.

We again start with the special case $p = q = 1$, where we also drop the subindex in the coefficient a. Then

$$x_{t|t-1} = -a x_{t-1} + E[u_t | X_{0,t-1}], \tag{9.138}$$

where $E[x|X]$ denotes the orthogonal projection of x on the space X. In the gaussian case this amounts to the conditional expectation, which explains the notation. The equations

$$x_1 = -ax_0 + u_1, \tag{9.139}$$

$$x_2 = -ax_1 + u_2, \tag{9.140}$$

$$\vdots$$

show that $X_{0,t} = V_{0,t}$, the space spanned by the sequence $v^t = x_0, u_1, \ldots, u_t$. Hence, $E[u_t|X_{0,t-1}] = E[u_t|V_{0,t-1}]$, which we still write as $u_{t|t-1}$. We orthogonalize this sequence by performing the Cholesky factorization of its covariance matrix $R(t) = E[v^t(v^t)']$ with the elements $r(0,0) = \sigma(0,0), r(0,1) = \sigma(0,1), r(t,t) = 1 + b^2$ and $r(t, t+1) = b$, the others being zero. The first two elements are the initial conditions. Writing this time $b(t,t) = b_0(t)$ and $b(t-1,t) = b_1(t)$, we get the representation:

$$x_0 = b_0(0)e_0, \tag{9.141}$$

$$u_1 = b_0(1)e_1 + b_1(1)e_0, \tag{9.142}$$

$$\vdots$$

$$u_t = b_0(t)e_t + b_1(t)e_{t-1}, \tag{9.143}$$

where e_t is an orthonormal sequence (not the same as in (9.137), though), and the coefficients are given by the factorization as

$$b_0(0) = +\sqrt{\sigma(0,0)}, b_1(1) = \sigma(0,1), \tag{9.144}$$

$$b_1(t) = b/b_0(t-1), \tag{9.145}$$

$$b_0(t) = [1 + b^2 - b_1^2(t)]^{1/2}, t > 0. \tag{9.146}$$

With these we then get

$$u_{t|t-1} = b_1(t)e_{t-1}, \tag{9.147}$$

$$u_t - u_{t|t-1} = b_0(t)e_t. \tag{9.148}$$

Hence, $x_{t|t-1} = -ax_{t-1} + u_{t|t-1}$ and $x_t - x_{t|t-1} = u_t - u_{t|t-1}$, which further gives $x_{t|t-1} = -ax_{t-1} + b_1(t)e_{t-1}$. By elimination of e_{t-1} we finally get

$$x_{t|t-1} + [b_1(t)/b_0(t-1)]x_{t-1|t-2} = [(b_1(t)/b_0(t-1)) - a]x_{t-1}, \tag{9.149}$$

which with the initial condition $x_{0|-1} = 0$ completes the solution.

In the same manner we can treat the general case. The Cholesky factorization gives us the representation of the process u as follows:

$$u_t = \epsilon_t + c_1(t)\epsilon_{t-1} + \cdots + c_q(t)\epsilon_{t-q}, \, t \geq q, \qquad (9.150)$$

where ϵ_t is an uncorrelated (but not of unit variance) process and

$$c_i(t) = b_i(t)b_0^{-1}(t-i), \, i = 1, \ldots, q. \qquad (9.151)$$

The representation (9.150) should be compared with the MA representation (9.137). Hence, even if we had started with the u-process (9.137), the fact that we predict from finitely many past observations forces us to create a different uncorrelated sequence, namely, (9.150). To write the recurrence equations for $b_i(t)$, note that the covariance matrix $R(t)$ consists of two kinds of elements: the initial ones $\sigma(i,j)$ and the rest, determined by the coefficients b_i. For the sake of uniformity, we extend the notation $r(i,j)$ to all the elements. Then with the slightly simpler notation $b_i(t) = b(t-i, t)$ we get

$$b_q(t) = r(t-q, t)b_0^{-1}(t-q), \qquad (9.152)$$

$$b_{q-1}(t) = [r(t-q+1, t) - b_q(t)b_1(t-q+1)]b_0^{-1}(t-q+1), \qquad (9.153)$$

$$\vdots$$

$$b_1(t) = [r(t-1, t) - b_q(t)b_{q-1}(t-1) - \cdots - b_2(t)b_1(t-1)]b_0^{-1}(t-1), \qquad (9.154)$$

$$b_0(t) = [r(t,t) - b_q(t)b_q(t) - \cdots - b_1(t)b_1(t)]^{1/2}, \, t > 0 \qquad (9.155)$$

$$b_0(0) = +\sqrt{r(0,0)}, \, b_i(t) = 0, \, i > t. \qquad (9.156)$$

From (9.150) we can finally write the predictor in the form

$$x_{t|t-1} + c_1(t)x_{t-1|t-2} + \cdots + c_q(t)x_{t-q|t-q-1} = d_1(t)x_{t-1} + \cdots + d_k(t)x_{t-k}, \qquad (9.157)$$

where $d_i(t) = c_i(t) - a_i$, $i = 1, \ldots, k$ for $k = \max\{p, q\}$, and the coefficients with undefined index values are zero.

One can show that if the u-process is stationary, say given by (9.137), and if $|R(t)| > \alpha > 0$, for all t, which is true if the polynomial defined by the coefficients in (9.137) has its roots strictly outside the unit circle, then $c_i(t) \to c_i$, and the polynomial $c(z) = 1 + c_1 z + \cdots + c_q z^q$ defines a stable system, i.e. has its roots strictly outside the unit circle. In fact, if the u-process was defined by (9.137), and the associated polynomial defines a stable system, then $c_i = b_i$. The limiting predictor then agrees with the stationary optimal predictor (9.132). Hence, the recurrence equations (9.157) provide one way to obtain the factorization of a strictly positive spectral matrix on the unit circle as

$$\Phi(z) = r_0 + r_1(z + z^{-1}) + \cdots + r_q(z^q + z^{-q}) = b(z)b(z^{-1}), \quad (9.158)$$

where, moreover, $b(z)$ has all its roots outside the unit circle.

9.4.2 Prediction bound with estimated parameters

A drawback of the preceding prediction formulas is that they require the knowledge of the ARMA parameters. We study in this subsection prediction when the parameters must be estimated. The prediction formulas above allow us to write a normal density function for an ARMA(p,q) model. Since the effect of the covariance of the initial conditions fades away at an exponential rate we use the simple predictor (9.132) and write $\theta = \mathbf{a}, \mathbf{b}, \sigma^2$, where $\mathbf{a} = a_1, \ldots, a_p$ and $\mathbf{b} = b_1, \ldots, b_q$ instead of the more involved one in (9.137). Consider the model

$$f(y^n; \theta) = \frac{1}{(2\pi\sigma^2)^{n/2}} e^{-\sum_t (y_t - y_{t|t-1})^2/(2\sigma^2)}, \quad (9.159)$$

where the mean values are given by the predictions

$$y_{t|t-1} + \sum_{i=1}^{q} b_i y_{t-i|t-i-1} = \sum_{i=1}^{q \wedge p} (b_i - a_i) y_{t-i}, \quad (9.160)$$

and where $q \wedge p$ is the maximum of p and q and the undefined components are zero together with $y_t = y_{t|t-1} = 0$ for $t < 1$. If the data are generated by the ARMA model

$$y_t + \sum_{i=1}^{p} a_i y_{t-i} = e_t + \sum_{i=1}^{q} b_i e_{t-i}, \quad t \geq 1, \quad (9.161)$$

for $\{e_t\}$ an iid gaussian $(0, \sigma^2)$ sequence, then $y_t - y_{t|t-1} = e_t$, and $f(y^n; \theta)$ is the product of the gaussian $(0, \sigma^2)$ random variables.

For any data sequence y^n the ML estimates $b_i(n)$ and $a_i(n)$ for each σ^2 minimize the sum

$$\min_{\mathbf{a}, \mathbf{b}} \sum_{t} (y_t - y_{t|t-1})^2, \quad (9.162)$$

and the minimized variance is given by

$$\hat{\sigma}^2 = \frac{1}{n}\sum_t (y_t - \bar{y}_t)^2, \qquad (9.163)$$

where

$$\bar{y}_t + \sum_{i=1}^{q} b_i(n)\bar{y}_{t-i} = \sum_{i=1}^{q \wedge p}(b_i(n) - a_i(n))y_{t-i}. \qquad (9.164)$$

There exists a large literature on numerical calculations of the ML estimates, but for our purpose it is enough to know what they are and that they can be calculated.

We derive a lower bound for all predictors as a corollary of the Theorem 5.3.

Theorem 9.3 *Let $\bar{y}_t = p(y^{t-1})$ be any predictor function. Then for all positive ϵ the inequality*

$$\frac{1}{n}\int f(y^n;\theta)\sum_{t=1}^{n}(y_t - \bar{y}_t)^2 dy^n \geq \sigma^2\left(1 + \frac{(p+q-\epsilon)}{n}\ln n\right) \qquad (9.165)$$

holds for n large enough and for all θ, except in a set whose volume goes to zero as n grows to infinity.

Proof Consider the density function defined by any predictor $\bar{y}^n$, not necessarily of type (9.164),

$$\bar{f}(y^n) = \frac{1}{(2\pi\sigma^2)^{n/2}} e^{-\sum_t (y_t - \bar{y}_t)^2/(2\sigma^2)}, \qquad (9.166)$$

and calculate the KL distance

$$E_\theta \ln \frac{f(y^n;\theta)}{\bar{f}(y^n)} = \frac{n}{2}\ln(2\pi\sigma^2) + E_\theta \frac{1}{2\sigma^2}\sum_t (y_t - \bar{y}_t)^2 - \frac{n}{2}\ln(2\pi e\sigma^2), \qquad (9.167)$$

where the last term is the entropy of the model $f(y^n;\theta)$ in (9.161). The condition in Theorem 5.3 holds, and $E_\theta \sum_t (y_t - \bar{y}_t)^2 \geq \sigma^2(n + \frac{1}{2}(p+q-\epsilon)\ln n)$ with the quantifications given in the theorem. The proof is complete.

The question arises of whether the predictor

$$\hat{y}_{t|t-1} + \sum_{i=1}^{q} b_i(t-1)\hat{y}_{t-i|t-i-1} = \sum_{i=1}^{q \wedge p}(b_i(t-1) - a_i(t-1))y_{t-i}, \qquad (9.168)$$

suggested by (9.160) will reach the lower bound. Whether it does depends on the asymptotic behavior of the sequential ML estimators $b_i(t-1)$ and $a_i(t-1)$

and that of the model

$$\hat{f}(y^n) = \frac{1}{(2\pi\sigma^2)^{n/2}} e^{-\sum_t (y_t - \hat{y}_{t|t-1})^2/(2\sigma^2)} \tag{9.169}$$

defined by the predictor (9.168) for the data generated by (9.161). For optimality it would have to be asymptotically as good as the NML model $\hat{f}(y^n; k)$, which appears to be the case although the proof looks cumbersome.

Appendix A Elements of algorithmic information

Although the material in this appendix is not used directly in the rest of the book, it has a profound importance in understanding the very essence of randomness and complexity, which are fundamental to probabilities and statistics. The algorithmic information or complexity theory is founded on the theory of recursive functions, the roots of which go back to the logicians Godel, Kleen, Church, and above all to Turing, who described a universal computer, the Turing machine, whose computing capability is no less than that of the latest supercomputers. The field of recursiveness has grown into an extensive branch of mathematics, of which we give just the very basics, which are relevant to statistics. For a comprehensive treatment we refer the reader to [26].

The field is somewhat peculiar in that the derivations and proofs of the basic results can be performed in two very different manners. First, since the partial recursive functions can be axiomatized as is common in other branches of mathematics the proofs are similar. But since the same set is also defined in terms of the computer, the Turing machine, a proof has to be a program. Now, the programs as binary strings of that machine are very primitive, like the machine language programs in modern computers, and they tend to be long and both hard to read and hard to understand. To shorten them and make them more comprehensible an appeal is often made to intuition, and the details are left to the reader to fill in. Having said this we give first the usual axioms without Turing machine interpretation so that at least we know what the important set of partial recursive functions is.

The objective is to describe functions from the set of integers N into the same set with operations as instructions, called *effective*, i.e. simple and unambiguous requiring no imagination so that even a machine can implement them. The domain can actually be any countable set, for we can easily map such sets to the integers in a one-to-one effective manner as seen below.

A.1 Recursive and partial recursive functions

The following list of six functions defines the *partial recursive functions* (prfs) starting with the more primitive so-called *recursive* functions (rfs):

1. Constant functions are recursive.
2. Successor function $S(n) = n + 1$ is recursive.
3. Projections $(x_1, \ldots, x_k) \mapsto x_i$ are recursive; so are compositions.
4. $g_1(x_1), \ldots, g_k(x_k)$ and $f(y_1, \ldots, y_k)$ recursive imply that $f(g_1(x_1), \ldots, g_k(x_k))$ is recursive.
5. f and g recursive imply that

$$h(0, x^k) = f(x^k), \tag{A.1}$$

$$h(1, x^k) = g(h(0, x^k), 1, x^k), \tag{A.2}$$

$$\vdots$$

$$h(n+1, x^k) = g(h(n, x^k), n, x^k) \tag{A.3}$$

is recursive.

6. All rfs are prfs. Then if $f(t, x_2, \ldots, x_k)$ is a prf, the crucial step is

$$g(x_1, x_2, \ldots, x_k) = \min_t [f(t, x_2, \ldots, x_k) = x_1] \tag{A.4}$$

is a prf.

Example The square root function is a prf

$$\sqrt{n} = \min_t [t^2 = n]. \tag{A.5}$$

A fundamental result is as follows. The set of prfs is the same as the set of functions f, where the calculation $n \mapsto f(n)$, if it exists, can be described by a computer program such as one for the Turing machine.

We list further important definitions:

1. A prf f is *total* and recursive, if its domain is N. (A total prf is not necessarily a recursive function.)

2. A real number is *computable* if there is a single program to compute each of its digits. Similarly, a function is computable if its values are computable.
3. A number or function is semicomputable from above (below) if there is a program to calculate converging approximations of it from above (below). However, we may never know how far from the limit the approximations are. A function is also called computable if it is semicomputable from both above and below (because now the approximation accuracy is known).

Generalization of the domain of a prf: Consider a recursive function $bin : B^* \leftrightarrow N$, where B^* denotes the set of all finite binary strings, defined by the pairs (x^n, m):

$$(\lambda, 0), (0, 1), (1, 2), (00, 3), (01, 4), (10, 5), (11, 6), \ldots, \tag{A.6}$$

sorted starting with the empty string λ, first by length and then lexically, [26]. Continue to a recursive function $N \times B^* \leftrightarrow B^*$, etc., which in a moment is seen to be particularly important.

Also the pairing function $pair : n \leftrightarrow (k, m)$ defined by the enumeration $n = 0, 1, 2, \ldots$ of $(0,0) \to (0,1) \to (1,0) \to (2,0) \to (1,1) \to (0,2) \to (0,3) \cdots$ in the two-dimensional array

$$
\begin{array}{cccc}
(0,0) \to (0,1) & (0,2) \to (0,3) \\
(1,0) & (1,1) & (1,2) \\
(2,0)
\end{array}
$$

is recursive. Repeating the simple rule of succession creates a sum $(m, n) \mapsto n + (m+n)(m+n+1)/2$ [26], which requires a more complicated property of the rf.

Further, a set $A \subset N$ (or $A \subset B^*$) is *recursive*, if the characteristic function is recursive:

$$n \in A \Rightarrow I_A(n) = 1,$$
$$n \notin A \Rightarrow I_A(n) = 0.$$

A set A is *recursively enumerable* (re) if A is empty, or it is the range of some total prf

$$f : N \to A \tag{A.7}$$

with repetitions allowed. In other words, A is re if some Turing machine or computer program can list its elements in some order with repetitions allowed

$$A = \{f(1), f(2), \ldots\}. \tag{A.8}$$

A.1.1 Universal computers

The interest in the set of prfs is that the set of the programs defining the set of prfs is itself re and hence admits a universal program that can list its elements. The universal program represents a finitely described "grammar" for the set of programs. Specifically, we have the following theorem.

Theorem A.1 *The set of prfs $\{f : B^* \to B^*\}$ is re, i.e. for some prf $F : N \times B^* \to B^*$ of arity 2*

$$F(i, x) = f_i(x), \tag{A.9}$$

for all i, whenever $f_i(x)$ exists.

$F(i, x)$ is *universal* in that it can implement any prf whose value exists. Also the set of prfs of arity 2 is re, and so on. If we view $i \leftrightarrow bin(i) \in B^*$ as a program for the function f_i, then $F : (bin(i), x) \mapsto f_i(x)$ is a universal computer for the set of prfs of arity 1 etc., which executes program $bin(i)$ that defines the output $f_i(x)$ for input x (if one exists).

Take a universal computer U and its programs $\{p_1, p_2, \ldots\}$ = language $\mathcal{L}$, sorted first by length and then lexically. The language is re. Also the set of all programs of arity 0, i.e. whose input is λ, and $U(p_i) = x$ for any string x is re by

$$F(i, \lambda) = p_i(\lambda) = x. \tag{A.10}$$

These ideas suggested the Solomonoff–Kolmogorov–Chaitin *complexity* (of the first original type):

$$K_U(x) = \min\{|p| : U(p) = x\}. \tag{A.11}$$

Let V be another universal computer with its programs $q_1, q_2, \ldots$ and universal function $G(i, \lambda)$. Then clearly

$$K_U(x) \leq K_V(x) + C_G. \tag{A.12}$$

Also

$$K_V(x) \leq K_U(x) + C_F \tag{A.13}$$

and

$$K_U(x) \doteq K_V(x) \tag{A.14}$$

to within a constant. We have an asymptotically absolute measure of complexity.

These ideas further suggest a definition of a universal probability measure for B^* by $2^{-K_U(x)}$, as originally desired by Solomonoff. This does not work, because the normalization

$$P_K(x) = \frac{2^{-K_U(x)}}{\sum_{y \in B^*} 2^{-K_U(y)}} \tag{A.15}$$

required is not finite.

The problem was fixed first by Kolmogorov and then even better by Chaitin [3] by restricting the programs to those that are self-delimiting. This makes the programs satisfy the Kraft inequality, and the normalizing coefficient does not exceed unity. However, we have to live with the fact that no program exists, which when given the string x will output the complexity nor of course calculate the normalizing coefficient. The usual proof of this important theorem due to Kolmogorov appeals to the undecidability of the so-called halting problem: No program exists which could always tell if a universal computer running a program will ever stop. Here is another proof, given by Ragnar Nohre [32].

Suppose to the contrary that a program q with the said property $q : x^n \mapsto K(x^n)$ exists. We can then write a program p, using q as a subroutine, which finds the shortest string z^n such that $K(z^n) > |p|$. In essence, the program p examines the strings x^t, sorted alphabetically by non-decreasing length, one after another, computes $K(x^t)$ with the subroutine q, and checks if this inequality holds. It is clear that such a shortest string exists, because $K(x^t)$ has no upper bound and $|p|$ has some fixed finite length. But by the definition of Kolmogorov complexity

$K(z^n) \leq |p|$, which contradicts the inequality shown. Hence, no program q with the assumed property exists.

A.1.2 Relative randomness and typicality

The asymptotic properties of Kolmogorov complexity have suggested opening the old problem of defining when a string could be considered "random." The intuitive idea is that if $K_U(x^n) \geq n$ we should consider the string *random*, because no regularities exist that could shorten the description of the string. There have been statistical ideas of infinite strings being random, which have met with difficulties. Martin-Lof [29] has made important contributions to removing some of these difficulties.

We are interested in the complexity based question of randomness of finite strings, which are relative to a fixed universal program and computer U. We wish to sharpen the quantification $K_U(x^n) \cong n$ in a manner that can also be applied to other than program defined models.

Since the length $|x|$ of a binary string x is determined by the string, consider the probability measure

$$P_S(x) = \hat{P}(x) = \frac{2^{-S_U(x)}}{\hat{C}_{|x|}}, \tag{A.16}$$

$$\hat{C}_{|x|} = \sum_{y \in B^{|x|}} 2^{-S_U(y)}, \tag{A.17}$$

where $S_U(x)$ is the shortest program length (A.11) without the prefix requirement, the letter S honoring Solomonoff, [54], [55] even though he overlooked the fact that the normalizing coefficient over the set of all strings is infinite, which was later fixed by Chaitin and Kolmogorov with the prefix requirement. We see that $\log \hat{C}_{|x|}$ plays the role of the maximum capacity of the model class defined by any set of programs generating the strings of length $|x|$. Indeed, we regard a set of programs $\mathcal{P} = \{p_y : y \in B^{|x|}\}$ as an "estimator" function $y \mapsto p_y$, which defines the distribution

$$\bar{P}(x) = \frac{2^{-|p_x|}}{\sum_{y \in B^{|x|}} 2^{-|p_y|}} = \frac{2^{-|p_x|}}{\bar{C}_{|x|}}. \tag{A.18}$$

The maximum of the normalizing coefficients $\bar{C}_{|x|}$ defines the maximum capacity $\log \hat{C}_{|x|}$. Notice that we have a code defined by the ideal lengths $\log 1/\hat{P}(x)$ for each length $|x| = n$, which is complete. The decoding can be done by taking the first string in growing lexical order such that

$$\min\{y^t : P_S(y^t) = P_S(x^n)\}. \tag{A.19}$$

It is easy to construct a prior for the lengths of the strings and get a single code for strings of all lengths, but for all priors the code is incomplete, and hence non-optimal.

By Theorem 4.1 the ideal code length $\log 1/\hat{P}(x)$ cannot be shortened by any means available in U, which we consider as the first requirement for randomness. In fact, the necessary condition for any program to be the shortest for a string x is that its length is $S_U(x)$. We add the second requirement of relative randomness and define a string x of length $|x|$ to be *random* relative to U, if

$$S_U(x) + \log \hat{C}_{|x|} \geq |x|. \tag{A.20}$$

Another notion of interest is the following: x is *typical* relative to any computable probability measure $P(x)$, if

$$S_U(x) \leq \log 1/P(x) \leq S_U(x) + \log \hat{C}_{|x|}. \tag{A.21}$$

Both of these notions will be further strengthened for hypothesis testing, because the parametric models we are interested in are far less powerful than the sets of computable distributions.

A.2 Kolmogorov structure function

We conclude this appendix with an outline of what is known as the Kolmogorov structure function. The algorithmic complexity, which originally was called "algorithmic information," has been criticized as being counterintuitive on the grounds that "random" strings, which have maximum complexity, have no "structure" or learnable information. Kolmogorov wanted to separate the amount of the learnable information, the "structure" of a string, from the overall complexity, defined by the Kolmogorov complexity; see [53]. As we shall see the procedure is nothing but an abstract version of the two-part MDL principle.

First, the Kolmogorov complexity is immediately extended to the conditional complexity $K(y|x)$ as the length of the shortest program that generates string y

given another string x and causes the computer to stop. One can show [26] that

$$K(x^n, y^n) < K(x^n) + K(y^n|x^n) + O(\log n). \tag{A.22}$$

Take $x = x^n$ to be a program that describes the "summarizing properties" of string $y = y^n$. A "property" of data may be formalized as a set A to which the data belong along with other sequences sharing this property. The number of properties is in inverse relation to the size of the set A. The smallest set is the singleton set $A = \{x^n\}$, and it represents *all* the conceivable properties of x^n, while the set consisting of all strings of length n captures no particular properties to x^n (other than the length n). We may now think of programs as consisting of two parts, where the first part describes optimally a set A with the number of bits given by the Kolmogorov complexity $K(A)$, and the second part describes x^n in A. Notice that the second part should not be measured by the conditional complexity $K(x^n|A)$, because the program might take advantage even of properties not in A. Rather, it is measured by $\log|A|$, where $|A|$ denotes the number of elements in A. There are other ways to measure this, some of which are described in [53]. The sequence x^n is then described in about $K(A) + \log|A|$ bits.

Consider the so-called *structure function*

$$h_{x^n}(\alpha) = \min_{A}\{\log|A| : x^n \in A, K(A) \leq \alpha\}. \tag{A.23}$$

Clearly, $\alpha < \alpha'$ implies $h_\alpha(x^n) \geq h_{\alpha'}(x^n)$ so that $h_{x^n}(\alpha)$ is a non-increasing function of α with the maximum value $n = \log 2^n$ at $\alpha = \log n$, or $\alpha = 0$ if n is given, and the minimum value $h_{x^n}(\alpha) = \log|\{x^n\}| = 0$ at $\alpha = K(x^n)$.

Because $\log|A| + K(A) \dot{\geq} K(x^n)$, where $\dot{\geq}$ denotes inequality to within a constant (similarly for $\dot{=}$), $h_{x^n}(\alpha)$ is above the line $L(\alpha) = K(x^n) - \alpha$ to within a constant. They are equal up to a constant at the smallest value $\alpha = \bar{\alpha}$ which satisfies

$$\min\{\alpha : h_{x^n}(\alpha) + \alpha \dot{=} K(x^n)\}, \tag{A.24}$$

and we get the Kolmogorov's *minimal sufficient statistic* decomposition of x^n:

$$\min_{A} h_{x^n}(K(A)) + K(A), \tag{A.25}$$

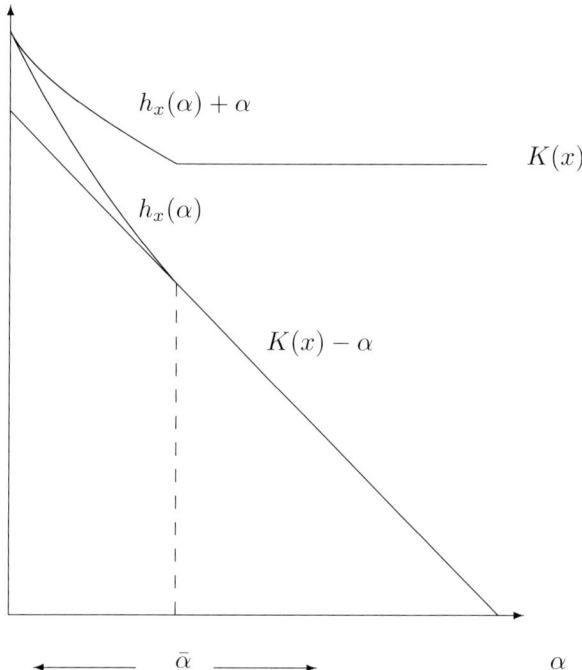

some properties captured   all properties captured

Figure A.1 The structure function and the optimal complexity $\bar{\alpha}$ of the model.

which amounts to the MDL principle applied to the two-part interval code length $L(x^n, A) = h_{x^n}(K(A)) + K(A)$. Once the properties have been extracted the remaining part represents the common worst case code length for each of the set of all "noise" sequences. Notice too that the structure function does not talk about "noise" as a sequence; it just gives its amount, measured in terms of code length. The problem of actually separating "noise" as a sequence from the data sequence belongs to "denoising."

Finally, while there is no point in talking about a "true" model A for the data, i.e. a "true" set A that includes the data, we can talk about the optimal model of complexity $\bar{\alpha}$, as illustrated in Figure A.1.

Appendix B Universal prior for integers

Especially in applications of the general MDL principle one frequently needs the code length for integers [10], or the easier-to-calculate negative logarithm of a universal prior for integers. For this reason we rederive a formula from [40].

Let $\log^k(n)$ denote the k-fold composition of the binary logarithm function, and let exp^k denote its inverse. For example, $exp^0(x) = x$, $exp^1(x) = 2^x$, $exp^2(2) = exp(exp(2)) = 2^4 = 16$, and $exp^3(2) = 2^{16}$. We also write $e(k) = exp^k(2)$. Further, let $log^*(n) = \log n + \log\log n + \cdots$, where the last term is the last positive k-fold composition. We wish to calculate the sum $c = \sum_{n>0} 2^{-\log^* n}$, more accurately than the upper bound given in [28].

The derivative of the k-fold logarithm function is given by

$$D\log^k(x) = 1/[a^k x \log x \times \cdots \times \log^{k-1}(x)] = a^{-k} 2^{-\log^* x}, \tag{B.1}$$

for $e(k-2) \leq x \leq e(k-1)$ and $e(i) = 0$ for $i < 0$, where $a = \ln 2 = 0.69...$. We can evaluate the integral

$$\int_{e(k-1)}^{e(k)} 2^{-\log^* x} dx = a^{k+1}, \tag{B.2}$$

which further gives

$$\int_{e(k)}^{\infty} 2^{-\log^* x} dx = a^{k+2}/(1-a) = S(k). \tag{B.3}$$

Next

$$2^{-\log^* n} < \int_{n-1}^{n} 2^{-\log^* x} dx < 2^{-\log^*(n-1)}, \tag{B.4}$$

which implies

$$\sum_{1}^{n-1} 2^{-\log^* i} + S(n) < c < \sum_{1}^{n-1} 2^{-\log^* i} + S(n-1). \tag{B.5}$$

The difference between the two bounds is by (B.4) less than $2^{-\log^*(n-1)}$. By putting $n - 1 = 16 = e(2)$ and calculating the upper bound from (B.5) we get $c = 2.86$ with an error less than 2^{-7}. By putting $n - 1 = 2^{16} = e(3)$ and calculating the upper bound from (B.5) we get $c = 2.865064$.

Put $w^*(n) = c^{-1} 2^{-\log^* n}$, which defines a *universal prior* for the integers. It gives the (ideal) code length

$$L^*(n) = \log 1/w^*(n) = \log^* n + \log c. \tag{B.6}$$

Let $L(n)$ be any length function satisfying the Kraft inequality. As in [4] consider the set of integers $A = \{n : L(n) \leq L^*(n) - k\}$, which is the same set as $\{n : w^*(n) \leq 2^{-L(n)-k}\}$, and let $I_A(n)$ be its indicator function, which is 1 if $n \in A$ and zero otherwise. Then

$$\sum_n w^*(n) I_A(n) \leq \sum_n I_A(n) 2^{-L(n)-k} \tag{B.7}$$

$$\leq \sum_n 2^{-L(n)-k} \leq 2^{-k}. \tag{B.8}$$

We see that the probability of the set of integers for which any prefix code length function can beat $L^*(n)$ by more than k bits decreases exponentially in k. This can be augmented by another result as follows. The uncertainty in the set of integers as measured by the entropy $\sum_n w^*(n) \log 1/w^*(n)$ is infinite. We prove the following theorem.

Theorem B.1 *Let the probability distribution P satisfy*

$$P(i) \geq P(i+1), \; i > M, \; some \; M, \tag{B.9}$$

$$H(P) = \infty. \tag{B.10}$$

Then

$$\lim_N \frac{\sum_{n \leq N} P(n) L^*(n)}{\sum_{n \leq N} P(n) \log 1/P(n)} = 1. \tag{B.11}$$

Proof Write $P_N = \sum_1^N P(i)$ and $L_N^* = \sum_1^N P(i) L^*(i)$. Similarly, let $H_N = \sum_1^N P(i) \log 1/P(i)$. By the noiseless coding theorem

$$\sum_{i=1}^N P(i) \log \frac{w_N^* P(i)}{P_N w^*(i)} = L_N^* - H_N + P_N \log(w_N^*/P_N) \geq 0. \tag{B.12}$$

Because $\epsilon \log \epsilon \to 0$, as $\epsilon \to 0$, the third term goes to zero as $N \to \infty$, and

$$\lim_N L_N^*/H_N \geq 1. \tag{B.13}$$

We have by Wyner's inequality [59] $j \geq 1/P(j)$

$$H_N \geq \sum_{j=1+M}^{N} P(j) \log 1/P(j) \geq \sum_{j=M+1}^{N} P(j) \log j. \tag{B.14}$$

Further,

$$L_N^* \leq \sum_{j=M+1}^{N} P(j) \log j + \sum_{j=M+1}^{N} P(j) r(j) + C, \tag{B.15}$$

where $r(j) = L^*(j) - \log j$, and C is the maximum of $L^*(j)$ for $j \leq M$. Let $f(j) = r(j)/\log j$, and let $K(\epsilon)$ be an index such that $K(\epsilon) > M$ and $f(j) \leq \epsilon$ for all $j \geq K(\epsilon)$. Then for $N > K(\epsilon)$,

$$\sum_{j=M+1}^{N} P(j) r(j) \leq \epsilon \sum_{j=K(\epsilon)+1}^{N} P(j) \log j + R(\epsilon), \tag{B.16}$$

where $R(\epsilon)$ denotes the maximum value of $r(i)$ for $i \leq K(\epsilon)$. With this inequality and (B.14) we get

$$L_N^*/H_N \leq 1 + \epsilon + [R(\epsilon) + C]/H_N. \tag{B.17}$$

This holds for all ϵ, and when $N \to \infty$ the theorem follows.

References

[1] Balasubramanian, V. (1996), Statistical inference, Occam's Razor and statistical mechanics on the space of probability distributions, *Neural Computation*, **9**, no. 2, 349–368, 1997. http://arxiv.org/list/nlin/9601.

[2] Barron, A.R., Rissanen, J., and Yu, B. (1998), The MDL principle in modeling and coding, *IEEE Trans. Information Theory* (special issue to commemorate 50 years of information theory), **IT-44**, no. 6, pp. 2743–2760.

[3] Chaitin, G.J. (1969), On the length of programs for computing finite binary sequences: statistical considerations, *JACM*, **16**, 145–159.

[4] Cover, T. and Thomas, J. (1991), *Elements of Information Theory*. New York: John Wiley and Sons, Inc.

[5] Cramér, H. (1946), *Mathematical Methods of Statistics*. Princeton: Princeton University Press.

[6] Csiszár, I. and Körner, J. (1981), *Information Theory: Coding Theorems for Discrete Memoryless Systems*. New York: Academic Press.

[7] Csiszár, I. and Shields, P. (2000), The consistency of BIC Markov order estimation, *Ann. Stat.*, **28**, no. 6, 1601–1619.

[8] Daubechies, I. (1988), Orthonormal bases of compactly supported wavelets, *Comm Pure and Appl. Math.*, **47**, no. 7, 909–996.

[9] Dawid, A.P. (1984), Present position and potential developments: some personal views, statistical theory, the prequential approach, *J. Royal Statist. Soc. A*, **147**, Part 2, 278–292.

[10] Elias, P. (1975), Universal codeword sets and representation of integers, *IEEE Trans. Information Theory*, **IT-21**, 194–203.

[11] Ferguson, T.S. (1996), *A Course in Large Sample Theory*, Texts in Statistical Science. London: Chapman and Hall.

[12] Fisher, R.A. (1925), Theory of statistical estimation, *Proc. Cambridge Phil. Soc.*, **22**, 700.

[13] Galves, A. and Leonardi, F. (2008), Exponential inequalities for empirical unbounded context trees, *Prog. Probability*, **60**, 257–270.

[14] Grünwald, P.D. (2007), *The Minimum Description Length Principle*. Cambridge: MIT Press.

[15] Hannan, E.J. and Quinn, B.G. (1979), The determination of the order of an autoregression, *J. R. Statist. Soc. B*, **41**, no. 2, pp. 190–195.

[16] Hansen, M.H. and Yu, B. (2001), Model selection and the Principle of Minimum Description Length, *J. Am. Statist. Assn*, **96**, no. 454, 746–774.

[17] Hartley, R.V. (1928), Transmission of information, *Bell System Technical J.*, **7**, 535–563.

[18] Huffman, D.A. (1952), A method for construction of minimum redundancy codes, *Proc. IRE*, **40**, 1098–1101.

[19] Kailath, T., Morf, M., and Sidhu, G.S. (1974), Some new algorithms for recursive estimation in constant discrete-time linear systems, *IEEE Trans. Automatic Control*, **AC-19**, 315–323.

[20] Kalman, R.E. (1960), A new approach to linear filtering and prediction problems, *Trans. ASME, J. Basic Engng.*, **82 D**, 34–45.

[21] Kolmogorov, A.N. (1965), Three approaches to the quantitative definition of information, *Problems of Information Transmission* **1**, 4–7.

[22] Kontkanen, P. and Myllymaki, P. (2007). A linear-time algorithm for computing the multinomial stochastic complexity, *Information Processing Lett.* **103**(6), 227–233.

[23] Krim, H. and Schick, I. (1999), Minmax descrption length for signal denoising and optimized representation, *IEEE Trans. Information Theory*, **45**, no. 3, 898–908.

[24] Lai, T.L. and Wei, C.Z. (1982), Least squares estimates in stochastic regression models with applications to identification and control of dynamic systems, *Ann. Stat.*, **10**, no. 1, 154–166.

[25] Lehmann, E.L. (2004), *Elements of Large-Sample Theory*. New York: Springer Verlag.

[26] Li, M. and Vitanyi, P.M.B. (1997), *An Introduction to Kolmogorov Complexity and its Applications*, second edition. New York: Springer Verlag.

[27] Lindquist, A. (1974), A new algorithm for optimal filtering of discrete-time stationary processes, *SIAM J. Control* **4**, 736–747.

[28] Leung-Yan-Cheong, S.K. and Cover, T. (1978), Some equivalences between Shannon enropy and Kolmogorov complexity, *IEEE Trans. Information Theory*, **24**, 331–338.

[29] Martin-Löf, P. (1966), On the concept of a random sequence, *Theory Prob. Applic.*, **11**, 177–179.

[30] Merhav, N. and Feder, M. (1995), A strong version of the redundancy-capacity theorem of universal coding, *IEEE Trans. Information Theory*, **IT-41**, no. 3, 714–722.

[31] Mononen, T. (2010), Computing the stochastic complexity of simple probabilistic graphical models, Ph.D. Thesis, Department of Computer Science, University of Helsinki, Finland.

[32] Nohre, R. (1993), Topics in descriptive complexity, Ph.D. thesis, Technical University of Linkoping, Sweden.

[33] Plackett, R.L. (1950), Some theorems in least squares, *Biometrika*, **37**, no. 1/2, 149–157.

[34] Rissanen, J. (1973), Algorithms for triangular decomposition of block Hankel and Toeplitz matrices with application to factoring positive matrix polynomials, *Math. Comput.*, **27**, 147–154.

[35] Rissanen, J. (1973), A fast algorithm for optimum linear predictors, *IEEE Trans. Automatic Control*, **AC-10**, 555.

[36] Rissanen, J. (1976), Generalized Kraft inequality and arithmetic coding, *IBM J. Res. Dev.*, **20**, 3, 198–203.

[37] Rissanen, J. (1978), Modeling by shortest data description, *Automatica*, **14**, 465–471.

[38] Rissanen, J. (1979), Arithmetic codings as number representations, *Acta Polytech. Scan.*, 44–51.

[39] Rissanen, J. (1983), A universal data compression system, *IEEE Trans. Information Theory*, **IT-29**, 5, 656–664.

[40] Rissanen, J. (1983), A universal prior for integers and estimation by minimum decsription length, *Ann. Stat.*, **11**, no. 2, 416–431.

[41] Rissanen, J. (1984), Universal coding, information, prediction, and estimation, *IEEE Trans. Information Theory*, **IT-30**, no. 4, 629–636.

[42] Rissanen, J. (1986), Stochastic complexity and modeling', *Ann. Statist.*, **14**, 1080–1100.

[43] Rissanen, J. (1992), Noise separation and MDL modeling of chaotic processes, pp. 317–330 in *From Statistical Physics to Statistical Inference and Back*, P. Grassberger and J.-P. Nadal (eds). The Netherlands: Kluwer Academic Publishers.

[44] Rissanen, J. (1996), Fisher information and stochastic complexity, *IEEE Trans. Information Theory*, **IT-42**, no. 1, 40–47.

[45] Rissanen, J. (2007), *Information and Complexity in Statistical Modeling*. New York: Springer Verlag.

[46] Rissanen, J., Roos, T., and Myllymaki, P. (2010), Model selection by sequentially normalized least squares, *J. Multivariate Anal.*, **101**, no. 4, 839849.

[47] Roos, T., Myllymaki, P., and Rissanen, J. (2009), MDL denoising revisited, *IEEE Trans. Signal Processing*, **57**, no. 9, 3347–3359.

[48] Shannon, C.E. (1948), A mathematical theory of communication, *Bell System Technical J.*, **27**, 379–423, 623–656.

[49] Shtarkov, Yu. M. (1987), Universal sequential coding of single messages, translated from *Problems of Information Transmission*, **23**, no. 3, 3–17, July–September 1987.

[50] Solomonoff, R.J. (1960), A preliminary report on a general theory of inductive inference', Report ZTB-135, Zator Co., Cambridge, MA, November 1960.

[51] Solomonoff, R.J. (1964), A formal theory of inductive inference, Part I, *Information and Control*, **7**, 1–22; Part II, *Inform. Control*, **7**, 224–254.

[52] Tabus, I. and Rissanen, J. (2006), Maximum likelihood model for logit regression, Festschrift for T. Pukkila, University of Tampere, Department of Mathematics, Statistics and Philosophy, Liski, E.P., Isotalo, J., Niemel, J., Puntanen, S., and Styan, G.P.H. (eds.), Department of Mathematics,

Statistics and Philosophy, Report A 368. ISBN 978-951-44-6620-5. pp. 295–300.

[53] Vereshchagin, N.K. and Vitanyi, P.M.B. (2004), 'Kolmogorov's Structure functions and model selection', *IEEE Trans. Information Theory*, **IT-50**, no. 12, 3265–3290.

[54] García, J.G. and González-López, V.A. (2010), Minimal Markov models, arXiv:1002.0729v1 [math.ST] 3 Feb 2010.

[55] Wald, A. (1945). Sequential tests of statistical hypotheses, *Ann. Mathemat. Stat.*, **16**, (2), 117–186.

[56] Wallace, C.S. and Boulton, D.M. (1968), An information measure for classification, *Computing J.*, **11**, no. 2, 185–195.

[57] Wei, C.Z. (1992), On predictive least squares principles, *Ann. Statist.*, **20**, no. 1, 1–42.

[58] Weinberger, M.J., Rissanen, J., and Feder, M. (1995), A universal finite memory source, *IEEE Trans. Information Theory*, **IT-41**, no. 3, 643–652.

[59] Wyner, A.D. (1972), An upper bound on entropy series, *Informat. Control*, **20**, 176–181.

Index

accept, 96, 99
alphabet, 11
atypical, 84

balanced tree, 14
batch, 50
Bayes, 13
Borel–Cantelli lemma, 66

capacity
   channel, 27
   maximum, 27
chi-square χ^2, 5
CLT, 5
code
   complete, 11, 13
   concatenation, 11
   mixture, 54
   prefix, 11
codeword, 11
complexity, 10, 14, 149
   algorithmic, 147
   stochastic, 53
compression, 49
computable, 146
confidence, 70
consistency, 63
   rate, 66
Cramér–Rao inequality, 57
critical region, 83

degree of belief, 13
distinguishability, 78

entropy, 1
   conditional, 24
estimation error, 77
estimator, 38
   maximum capacity (MC), 4, 43
   maximum likelihood (ML), 42
explanatory data, 35

Fisher information matrix, 53

hyperparameters, 6
hypothesis, 84
   composite, 84
   simple, 84

innovation, 136

KL distance, 16, 55

likelihood, 38
likelihood ratio, 102

maximum capacity (MC), 43, 149
maximum estimation information, 28
MC estimator, 46
MC interval estimator, 77
MDL principle
   complete, 51
   general, 51
mean information, 15
mean separation, 77
message, 11
model, 35
   class, 36
   NML, 52
   non-parametric, 37
   parametric, 36
   selection, 37
   structure, 36
multinomial, 5
mutual information, 25

noise, 86
null hypothesis, 83

orthonormal, 104, 135

perfectly fair, 50
precision, 70

radix
   double, 20
   single, 19

Index

random, 85, 149, 150
ratio test, 96
region
  acceptance, 97
  rejection, 97
regret, 56
rejection, 97, 99

semicomputable, 146
SNML, 114

source, 2
statistical significance, 83
symbols, 11

test statistic, 83
typical, 84, 150

unimodal, 89

wavelets, 104